MCR 2009

ADVANCES IN EXPERIMENTAL MEDICINE AND BIOLOGY

For other titles published in this series, go to
www.springer.com/series/5584

Maxim A. Mironov
Editor

MCR 2009

Proceedings of the 4th International Conference on Multi-Component Reactions and Related Chemistry, Ekaterinburg, Russia

Springer

Editor
Maxim A. Mironov
Department of Technology of Organic Synthesis
Ural State Technical University
Mira Str. 19
620002 Ekaterinburg
Russia
mcr2009@mail.ustu.ru

ISBN 978-1-4419-7269-9 e-ISBN 978-1-4419-7270-5
DOI 10.1007/978-1-4419-7270-5
Springer New York Dordrecht Heidelberg London

Library of Congress Control Number: 2010938603

Printed on acid-free paper

Springer is part of Springer Science+Business Media (www.springer.com)

Contents

Introduction to the Proceedings of the Fourth International Conference on Multi-Component Reactions and Related Chemistry

Maxim A. Mironov

The Fourth International Conference on Multi-Component Reactions and Related Chemistry was held in Ekaterinburg, Russia, from May 24 to 28, 2009. The conference evolved from a great tradition that was started by Prof. Ivar Ugi, who can be considered the father of the concept of Multi-Component Reactions (MCRs) and the grandfather of combinatorial chemistry. This meeting was a continuation of three highly successful conferences on the same topic that were held in Munich in 2000, in Genova in 2003, and in Amsterdam in 2006. All of the conferences had about 250 participants, about half of them coming from private companies. The quality and variety of the presented lectures greatly contributed to this outcome. Due to its unique geographical position on the border of Europe and Asia, Ekaterinburg assumes the role of an important center of cooperation between the East and the West. This place was ideal for the Fourth International Conference on Multi-Component Reactions and Related Chemistry, chaired by Prof. Oleg Chupakhin and Dr. Maxim Mironov, where 134 scientists from 17 countries covering Europe, North America, South Asia, and the Pacific regions discussed recent progress and further developments in the field. A total of 43 oral presentations and 52 posters were delivered. Our event was supported by two leader companies in pharma research: ChemDiv Inc. and Abbott Laboratories, and also by the Russian Academy of Science and Springer Science+Business Media. Materials of the conference were published in a special issue of *Molecular Diversity* (Springer). The conference was accompanied by an exhibition and training courses, which featured participation from the pharma industry, analytical equipment providers, and publishers.

The scientific program of the conference focused on MCRs, which are a rapidly expanding research area in medicinal chemistry and materials science. The classical MCR chemistry is well-known to those who have studied organic chemistry at the university level. Many of these classical reactions are named reactions, such as the Mannich reaction and its intramolecular variant, the Pictet–Spengler reaction, the Strecker synthesis, the Ugi four-component condensation, and many others. Nowadays, the number of research groups active in the MCR field is growing

M.A. Mironov (✉)
Urals State Technical University, Ekaterinburg, Russia
e-mail: mironov@mail.ustu.ru

rapidly and novel MCRs are being discovered or developed on a weekly basis. The number of publications in the MCR area has increased dramatically, reaching several hundred articles per year. More and more companies continue to adopt this chemistry in their synthetic repertoire. There are several advantages of synthesis using MCRs. First, they are fully amenable to automated synthesis because there is no need to isolate any intermediates and the experiments are carried out in a single reaction vessel. Second, MCRs are more convergent than synthesis pathways based on a sequence of uni- and bimolecular reactions. In addition, MCRs are a particular case of reaction networks, which can be considered a model of natural processes and prebiotic evolution. Although the main focus of the conference was on MCRs, its scope embraced organic synthesis via tandem or cascade reactions, combinatorial approaches to new materials, catalysts, and supramolecular structures. The conference was organized in six sections: (1) Catalysis and Multi-Component Reactions; (2) Multi-Component Reactions in Heterocyclic Chemistry; (3) Multi-Component Reactions in Drug Discovery; (4) Novel Reagents for Multi-Component Reactions; (5) Design of Multi-Component Reactions; and (6) Multi-Component Reactions in Supramolecular Chemistry and Material Science.

Organizing Committee:

Chairman

Prof. Oleg N. Chupakhin *(Russian Academy of Science, Russia)*

Vice-Chairman

Dr. Maxim A. Mironov *(Ural State Technical University, Ekaterinburg, Russia)*

International Scientific Committee

Prof. Luca Banfi *(University of Genova, Italy)*

Prof. Valery N. Charushin *(Russian Academy of Science, Russia)*

Prof. Alexander Dömling *(Pittsburgh University, USA)*

Prof. Romano Orru *(Vrije University, Amsterdam, The Netherlands)*

Prof. Bing Yan *(Shandong University, P.R. China)*

Local Organizing Committee

Prof. Vasiliy A. Bakulev *(Ural State Technical University, Ekaterinburg, Russia)*

Dr. Dmitry V. Kravchenko *(ChemDiv Inc., Moscow, Russia)*

Prof. Anatoly I. Matern *(Ural State Technical University, Ekaterinburg, Russia)*

Prof. Valentine G. Nenajdenko *(Moscow State University, Moscow, Russia)*

Prof. Vladimir L. Rusinov *(Ural State Technical University, Ekaterinburg, Russia)*

Dr. Elena V. Sadchikova *(Ural State Technical University, Ekaterinburg, Russia)*

Session 1
Catalysis and Multi-Component Reactions

Recent Progress in Asymmetric Two-Center Catalysis

Masakatsu Shibasaki

We have been studying the development of new asymmetric two-center catalysis using rare earth alkoxides and bifunctional sugar and related ligands. In The Fourth International Conference on Multi-Component Reactions and Related Chemistry (MCR 2009), new catalytic asymmetric reactions using catalysts **1** and **2** and catalytic asymmetric syntheses of ranirestat **3** and tamiflu **4** will be presented.

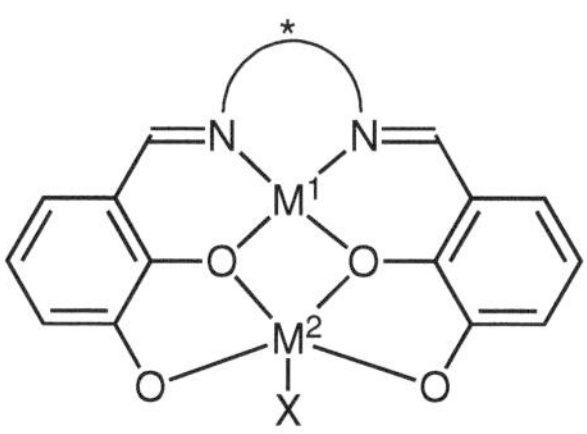

M^1-M^2-Schiff base **1** complex

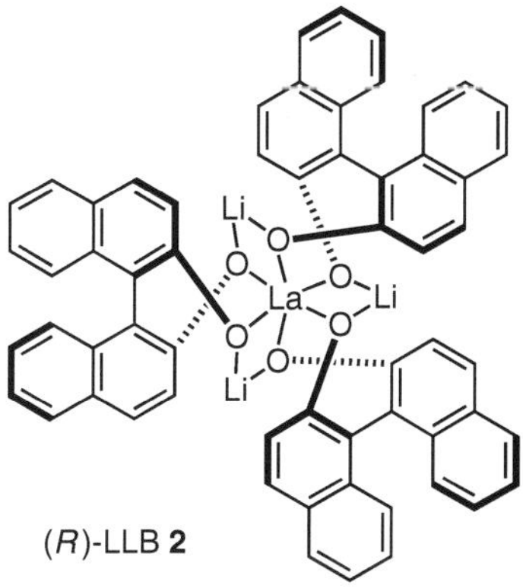

(*R*)-LLB **2**

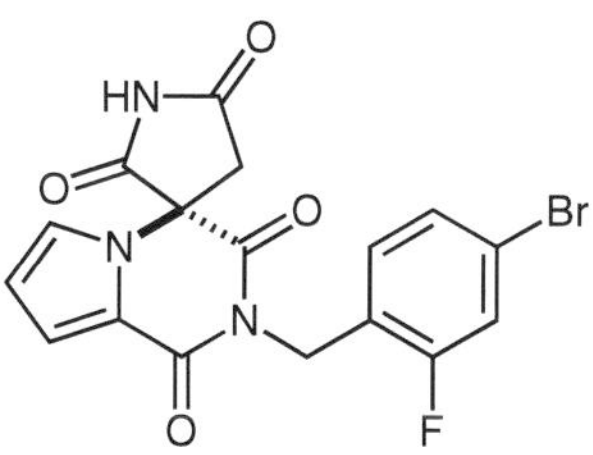

ranirestat **3**

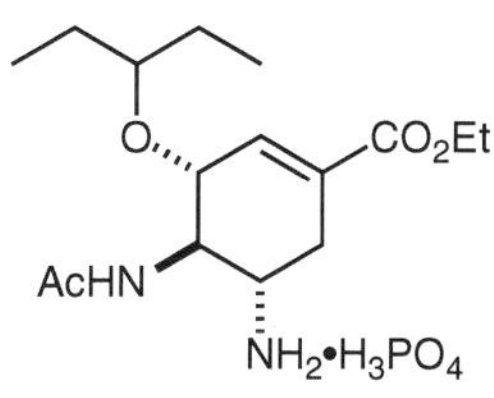

tamiflu **4**

M. Shibasaki (✉)
Graduate School of Pharmaceutical Sciences, University of Tokyo, Hongo 7-3-1, Bunkyoku, Tokyo 113-0033, Japan
e-mail: ysuya@mol.f.u-tokyo.ac.jp

M.A. Mironov (ed.), *MCR 2009*, Advances in Experimental Medicine and Biology 699, DOI 10.1007/978-1-4419-7270-5_1, © Springer Science+Business Media, LLC 2011

Reference

Oisaki K, Zhao DB, Kanai M, Shibasaki M. (2007) Catalytic enantioselective alkylative aldol reaction: efficient multicomponent assembly of dialkylzincs, allenic esters, and ketones toward highly functionalized delta-lactones with tetrasubstituted chiral centers. J Am Chem Soc 129:7439–7443.

Efficiency in Chemistry: From Hydrogen Autotransfer to Multi-component Catalysts

Miguel Yus

One important task concerning any chemical process has to do with the so-called atomic efficiency (AE), which can considerably modify the concept of yield corresponding to a chemical reaction: Even working with a high chemical yield a reaction can be inefficient when the main part of the reactant's structure is not included in the final product. Two interesting processes will be the subject of this presentation: (a) the hydrogen autotransfer reaction, in which an alcohol is used as the electrophilic component in the alkylation of a carbonyl compound; water is the only byproduct in the process, which is, therefore, of great value from an atom-efficiency point of view (1); and (b) the magnetite-catalyzed multi-component aza-Sakurai reaction, of considerable interest from a synthetic efficiency point of view (2).

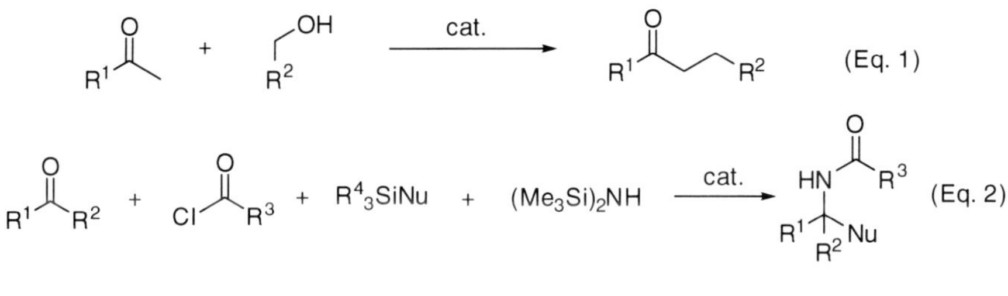

References

Guillena G, Ramón DJ, Yus M. (2007) Alcohols as electrophiles in C–C bond-forming reactions: the hydrogen autotransfer process. Angew Chem Int Ed 46:2358–2364.

Ramón DJ, Yus M. (2005) Asymmetric multicomponent reactions (AMCRs): the new frontier. Angew Chem Int Ed 44:1602–1634.

M. Yus (✉)
Institute for Organic Synthesis and Department of Organic Chemistry, Faculty of Science, University of Alicante, Apdo. 99, 03080 Alicante, Spain
e-mail: yus@ua.es

Multi-component Reactions by Microwave-Assisted, Continuous Flow Organic Synthesis

Stacy Bremner, Eamon Comer, Gjergji Shore, Sylvie Morin, and Michael G. Organ

A new technology Microwave-Assisted, Continuous Flow Organic Synthesis (MACOS) for the synthesis of compounds in flow has been developed that uses microwave irradiation to promote the chemical transformation (Fig. 1).

Further, it has been discovered that metal films that are laid down on the walls of these microscale flow tubes (capillaries) couple very effectively with microwave irradiation to drive a variety of reactions to completion very rapidly. Routine processes have been developed for the deposition of any metal onto the inner surface of micro-capillaries, and these films have been characterized by Scanning Electron Microscopy (SEM) and Energy Dispersive X-ray (EDX). These films have been shown to be highly catalytically active in metal-catalyzed transformations including the Suzuki–Miyama and Heck couplings, hydrosilylation, pericyclic reactions, alkyne additions to imines, and click chemistry.

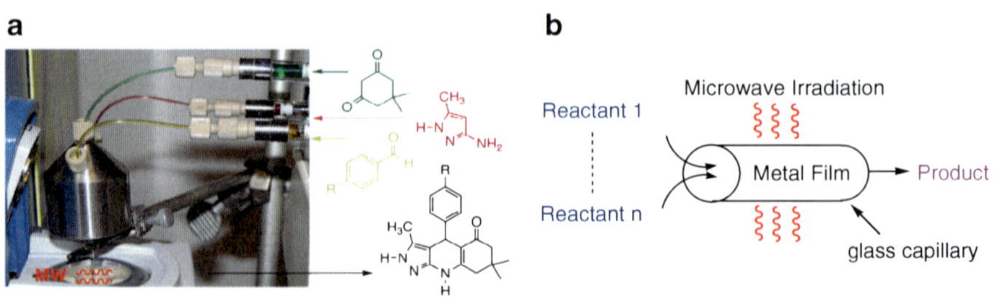

Fig. 1 (**a**) Photograph of a MACOS reactor design configured for three-component MCR chemistry. (**b**) Schematic of MACOS setup involving a metal-coated glass capillary

M.G. Organ (✉)
Department of Chemistry, York University, Toronto, ON, M3J 1P3, Canada
e-mail: organ@yorku.ca

In this presentation, applications of MACOS and the above-described metal-film technology will be discussed in terms of their application to MCR chemistry. Further, aspects pertaining to "on-the-fly" in-line analysis, reaction optimization, and the scale-up of chemical reactions will be touched upon.

References

Shore G, Organ MG. (2008) Gold-film-catalyzed hydrosilylation of alkynes by microwave-assisted, continuous-flow organic synthesis (MACOS). Chem Eur J 14:9641–9646.

Shore G, Organ MG. (2008) Diels–Alder cycloadditions by microwave-assisted, continuous flow organic synthesis (MACOS): the role of metal films in the flow tube. Chem Commun 838–840.

Shore G, Morin S, Mallik D, Organ MG. (2008) Pd PEPPSI-IPr-mediated reactions in metal-coated capillaries under MACOS: the synthesis of indoles by sequential aryl amination/Heck coupling. Chem Eur J 14:1351–1356.

A Potpourri of Recent Microwave-Assisted 2(1*H*)-Pyrazinone Chemistry

Erik Van der Eycken

In the last 3 decades, 3,5-dichloro-2(1*H*)-pyrazinones have emerged as useful starting materials for the elaboration of different types of skeletons of biologically interesting compounds. The 2(1*H*)-pyrazinone scaffold allows the easy introduction of a wide range of pharmacologically active groups with the ability to address a diverse set of biological targets. We will comment on our latest results regarding the application of focused microwave irradiation for the decoration and conversion of this useful scaffold. Our recent results about the first palladium-catalyzed desulfitative Sonogashira-type cross-coupling reaction as well as those concerning a desulfitative Hiyama-type cross-coupling will be presented. We will also comment on the development of a novel and versatile entry to asymmetrically substituted pyrazines, including a microwave-assisted Liebeskind–Srogl protocol, as well as on the elaboration of an unprecedented route for the synthesis of dihydropyrazine-2, 3-diones applying aqueous ("green") conditions. A highly efficient method for the diversity-oriented synthesis of tri- and tetrasubstituted furo[2,3-b]pyrazines has been developed comprising an Ag^+- or iodine-mediated intramolecular heteroannulation reaction.

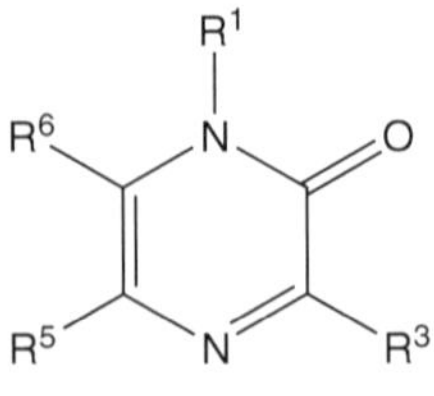

The 2(1*H*)-pyrazinone scaffold

E. Van der Eycken (✉)
Laboratory for Organic & Microwave-Assisted Chemistry (LOMAC), Katholieke Universiteit Leuven (K.U. Leuven), Celestijnenlaan 200F, B-3001, Leuven, Belgium
e-mail: erik.vandereycken@chem.kuleuven.be

References

Koval N, Appukkuttan P, Van der Eycken E. (2006) The chemistry of 2-(1*H*)-pyrazinones in solution and on solid support. In: Van der Eycken E, Kappe O (Eds.), Topics in Heterocyclic Chemistry, Microwave-Assisted Synthesis of Heterocycles. Springer, Berlin-Heidelberg-New York.

Mehta VP, Sharma A, Van der Eycken E. (2008) The first palladium-catalyzed desulfitative Sonogashira-type cross-coupling of (hetero)aryl thioethers with terminal Alkynes. Org Lett 10:1147–1150.

Mehta VP, Sharma A, Van Hecke K, Van Meervelt L, Van der Eycken E. (2008) A novel and versatile entry to asymmetrically substituted pyrazines. J Org Chem 73:2382–2388.

Sharma A, Mehta VP, Van Hecke K, Van Meervelt L, Van der Eycken E. (2008) Synthesis of 5-(phenylsulfanyl)-1,4-dihydropyrazine-2,3-diones via an unexpected microwave-assisted cascade reaction. Tetrahedron Lett 49:4993–4996.

Efficient Construction of (Hetero)Aryl Substituted (Aza)Indoles by Pd-Catalyzed Multi-component Reactions

Eugen Merkul, Dieter Dorsch, and Thomas J.J. Müller

Indoles and 7-azaindoles bearing heterocyclic substituents in the 3-position constitute a very important class of natural alkaloids and synthetic biologically active compounds. Pd-catalyzed cross-coupling methodologies are reliable and efficient tools for C–C bond formation. Therefore, our strategy for the introduction of various heterocycles in the 3-position of (aza)indoles is based on mild and efficient protocols derived from Sonogashira ynone synthesis, Suzuki biaryl coupling, and Masuda borylation. The installation of pyrimidyl substituents can be performed by two complementary multi-component approaches. The carbonylative alkynylation under one atmosphere of carbon monoxide and unprecedented decarbonylative *Sonogashira* coupling provide efficient accesses to indolyl ynones. These useful intermediates are then converted into pyrimidines by classical cyclocondensation. Furthermore, a sequence of Masuda and Suzuki cross-couplings allows us to decorate (aza)indoles with diverse heterocyclic substituents in the 3-position using the same Pd-catalyst for sequential transformations in a one-pot fashion.

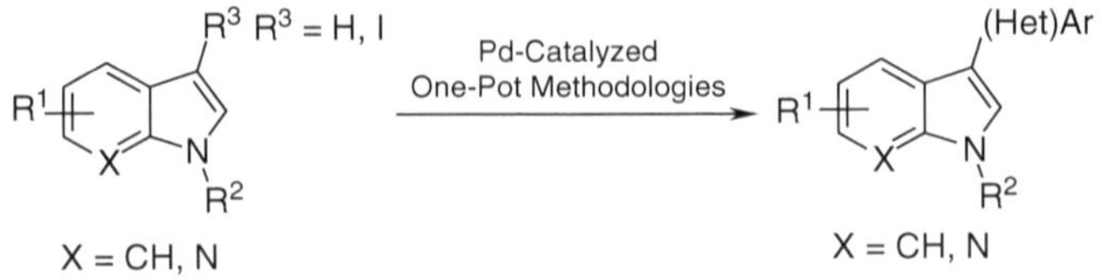

The strategy, results, and some biological data of targets will be presented.

E. Merkul (✉)
Institut für Organische Chemie und Makromolekulare Chemie, Heinrich-Heine-Universität Düsseldorf, Universitätsstrasse, 1, 40225, Düsseldorf, Germany
e-mail: E.Merkul@uni-duesseldorf.de

References

Karpov AS, Müller TJJ. (2003) New entry to a three-component pyrimidine synthesis by TMS–Ynones via Sonogashira coupling. Org Lett 5:3451.

Murata M, Watanabe S, Masuda Y. (1997) Novel palladium(0)-catalyzed coupling reaction of dialkoxyborane with aryl halides: convenient synthetic route to arylboronates. J Org Chem 62:6458.

Karpov AS, Merkul E, Rominger F, Müller TJJ. (2005) Concise syntheses of meridianins by carbonylative alkynylation and a four-component pyrimidine synthesis. Angew Chem Int Ed 44:6951.

Chlorotrimethylsilane is an Efficient Promoter for the Biginelli Reaction

Sergey V. Ryabukhin, Eugeniy N. Ostapchuk, Andrey S. Plaskon, Dmitriy M. Volochnyuk, Alexander N. Shivanyuk, and Andrey A. Tolmachev

The Biginelli reaction is typically the acid-catalyzed one-pot multi-component cyclocondensation of an aldehyde, a β-ketoester, and a urea, resulting in various dihydropyrimidines. The latter have shown to be efficient calcium channel modulators, mitotic kinesine inhibitors, adrenergic receptor antagonists, and antibacterial and antiviral agents. Such a wide spectrum of biological activity allows consideration of the dihydropyrimidine structural unit as one of the most important drug-like scaffolds.

We discovered a new, efficient methodology for the Biginelli reaction which has many possibilities for widening its scope and limitation. A diverse set of all components is examined in the reaction promoted by chlorotrimethylsilane.

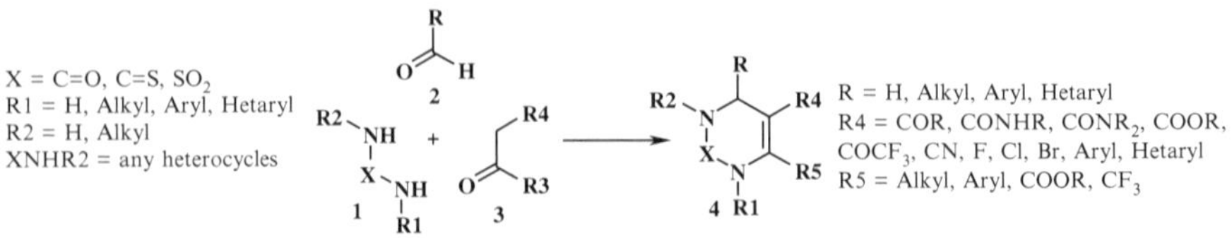

Also, some new Biginelli-type multi-component reactions have been developed.

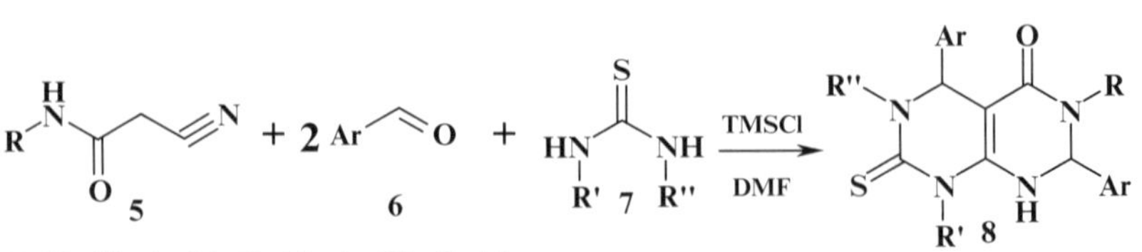

R=H, Alk, Ar; R'= H, Alk, Ar; R"= H, Alk

S.V. Ryabukhin (✉)
Enamine Ltd, Alexandra Matrosova Street 23, Kiev, 01103, Ukraine
e-mail: ryabukhin@mail.enamin.net

References

Ryabukhin SV, Plaskon AS, Ostapchuk EN, Volochnyuk DM, Tolmachev AA. (2007) N-Substituted ureas and thioureas in Biginelli reaction promoted by chlorotrimethylsilane: convenient synthesis of N1-alkyl-, N1-aryl-, and N1,N3-dialkyl-3,4-dihydropyrimidin-2(1H)-(thi)ones. Synthesis 3:417–427.

Ryabukhin SV, Plaskon AS, Ostapchuk EN, Volochnyuk DM, Shishkin OV, Shivanyuk AN, Tolmachev AA. (2007) A one-step fusion of 1,3-thiazine and pyrimidine cycles. Org Lett 21:4215–4218.

Three-Component Synthesis of Dihydropyrimidobenzimidazol Derivatives in the Presence of MgO as a Highly Effective Heterogeneous Base Catalyst

Hassan Sheibani, Maryam Babaie, and Soheila Behzadi

Condensed pyrimidine compounds with a five-membered aromatic ring consisting of two heteroatoms such as benzimidazoles or benzothiazoles have been shown to exhibit interesting pharmacological properties, and a number of synthetic methods have been reported for their preparation. As part of our current studies on the development of new routes to the synthesis of heterocyclic compounds, herein we now report a new and efficient method for the preparation of 2-amino-4-aryl-4, 10-dihydropyrimido[1,2-*a*][1,3]benzimidazol-3-yl cyanide **4a** or 2-amino-4-aryl-4H-pyrimido[2,1-*b*][1,3]benzothiazol-3-yl cyanide **4b** derivatives from the three-component reaction of aromatic aldehydes **2**, malononitrile **3**, and 2-aminobenzimidazoles **1a** or 2-aminobenzothiazoles **1b**, respectively, in the presence of base catalysts such as sodium acetate or high-surface-area MgO as a highly effective heterogeneous base catalyst in excellent yields and a short experimental time.

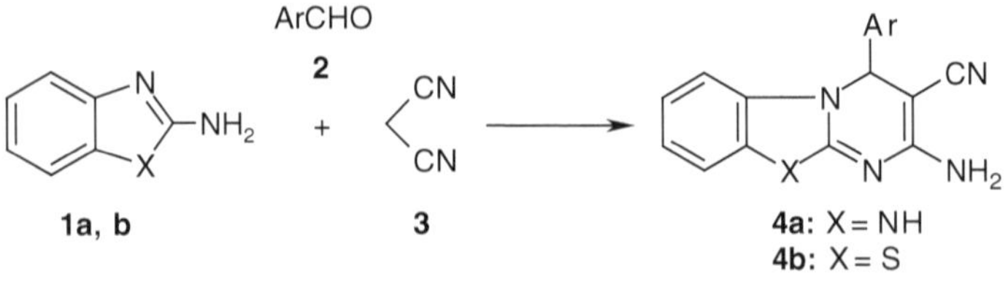

The structure of compounds **4a**, **b** was determined on the basis of their elemental analyses, mass spectrometry, ^{1}H and ^{13}C NMR and IR spectroscopic data.

H. Sheibani (✉)
Department of Chemistry, Shahid Bahonar University of Kerman, Kerman 76169, Iran
e-mail: hsheibani@mail.uk.ac.ir

References

Landreau C, Deniaud D, Evain M, Reliquet A, Meslin JC. (2002) Efficient regioselective synthesis of triheterocyclic compounds: imidazo[2,1-*b*]benzothiazoles, pyrimido[2,1-*b*]benzothiazolones and pyrimido[2,1-*b*]benzothiazoles. J Chem Soc Perkin Trans 1:741.

Sheibani H, Zahedifar M. (2009) The condensation of (chlorocarbonyl)phenylketene with 1,3-dinucleophiles. II. Preparation of 2-hydroxy-3-phenyl-4H-pyrimido[2,1-b][1,3]benzothiazol-4-ones and thioxo dihydro-4,6(1H,5H)-pyrimidinones. Heterocycles 78:748–753.

[Hmim]TFA-Catalyzed Multi-component Reaction: Direct, Mild, and Efficient Procedure for the Synthesis of 1,2-Dihydroquinazolines

Minoo Dabiri, Mahboobeh Bahramnejad, and Sahareh Bashiribod

4-Substituted-spiro-1,2-dihydroquinazolines and the related compounds are synthesized by direct reaction of 2-aminobenzophenones, isatin derivatives, and ammonium acetate in the presence of the dual-role catalyst-solvent ionic liquid, [Hmim]TFA. An excellent conversion of starting materials is achieved to the desired dihydroquinazoline products.

The multi-component reaction is one of the most efficient synthetic methods for organic molecules. The 1,2-dihydroquinazoline family of compounds has a rich pharmacology with reported nitric oxide synthase (NOS) inhibitors and antiinflammatory efficiency. Herein we disclose a synthetic route that allows access to high yields of a variety of dihydroquinazoline derivatives. In view of the emerging importance of imidazolium-based ionic liquids as novel reaction media, we also explore the use of methylimidazolium trifluoroacetate, [Hmim]TFA, as a promoter solvent and acidic catalyst for preparation of dihydroquinazolines under mild conditions. The synthesis of these compounds is now more convenient, clean, and efficient. There is no need for the use of polar aprotic solvents such as DMF, or solvents in general, and minimal waste is generated.

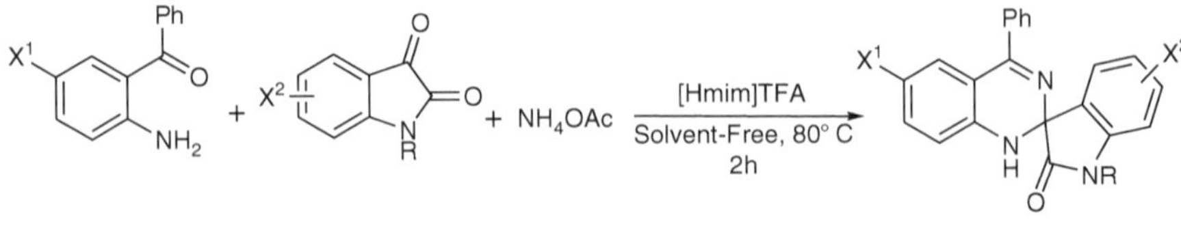

M. Dabiri (✉)
Department of Chemistry, Faculty of Science, Shahid Beheshti University, G. C., Tehran 1983963113, Iran
e-mail: m-dabiri@sbu.ac.ir

References

Dömling A, Ugi I. (2000) Multicomponent reactions with isocyanides. Angew Chem Int Ed 39:3168.

Tinker AC, Beaton HG, Smit NB, Cook TR, Cooper SL, Rae LF, Hallam K, Hamley P, McInally T, Nicholls DJ, Pimm AD, Wallace AV. (2003) 1,2-Dihydro-4-quinazolinamines: potent, highly selective inhibitors of inducible nitric oxide synthase which show antiinflammatory activity *in vivo*. J Med Chem 46:913.

Shen ZL, Ji SJ, Loh TP. (2005) Ionic liquid [omim][PF_6] as an efficient and recyclable reaction media for the cyanosilylation of aldehydes without Lewis acid or any special activation. Tetrahedron Lett 46:3137.

Synthesis of Benzofurans Catalyzed by 1,4-Diazabicyclo[2,2,2]Octane in Solventless System

Rahim Hekmatshoar, Sodeh Sadjadi, and Mohammad Khorasani

Among nature's collection of biologically active heterocycles, benzofuran derivatives constitute a major group. They are usually an important constituent of plant extracts used in medicinal chemistry for their various biological activities, including insecticidal, traditional medicine, antimicrobial, and antioxidant properties. Despite the availability of multi-step methods for the synthesis of benzofurans, a straightforward approach for their synthesis from easily available starting materials is always advantageous.

In this work several benzofurans were prepared under solvent-free condition at room temperature with good to excellent yields by simply grinding the mixture of relevant salicylaldehydes, phenacyl bromide, anhydrous potassium carbonate, and 1,4-diazabicyclo[2,2,2]-octane (DABCO) as catalyst in a mortar.

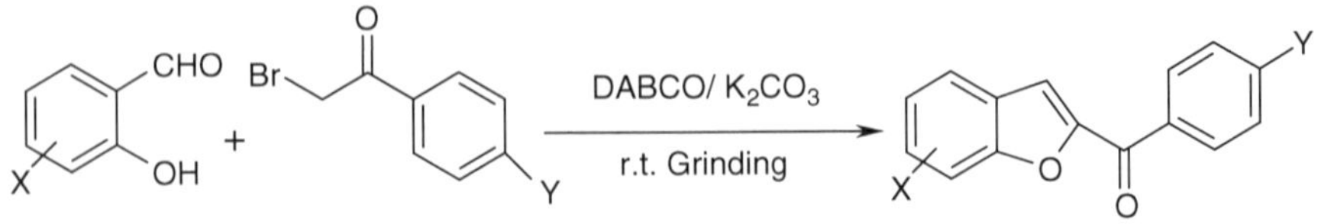

Reference

Herndon JW, Zhang Y, Wang H, Wang K. (2000) Synthesis of benzofuran derivatives through the coupling of conjugated dienynes with Fischer carbene complexes. Tetrahedron Lett 41:8687–8690.

R. Hekmatshoar (✉)
D epartment of Chemistry, School of Science, Alzahra University, Vanak, Tehran, Iran
e-mail: rhekmatus@yahoo.com

Silica Sulfuric Acid: Highly Efficient Catalyst for the Synthesis of Imidazo[1,2-a]pyridines Using Trimethylsilylcyanide or Cyanohydrins

Anatoliy I. Polyakov, Vera A. Eryomina, Lidiya A. Medvedeva, Nadezhda I. Tihonova, Anna V. Listratova, and Leonid G. Voskressensky

The current study was aimed at elaborating a synthetic protocol toward imidazo[1,2-*a*](di)azines avoiding the use of (1) isonitriles as a starting material, (2) costly catalysts (i.e., scandium triflate), and (3) special instrumentation (i.e., a microwave reactor). Having this idea in mind, we turned our attention to the use of an H_2SO_4/SiO_2 catalytic system, known as silica sulfuric acid. TMSCN or cyanohydrins of acetone or acetaldehyde were used instead of isonitriles.

We have worked out the method for the synthesis of 3-aminoimidazole using silica sulfuric acid catalyst for the condensation of aminoazines and cyano-substituted

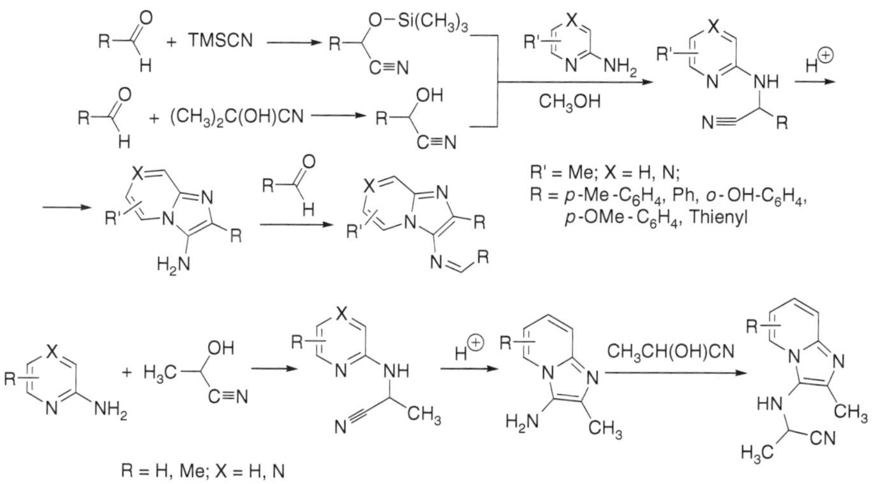

Scheme 1 Synthesis of 3-aminoimidazoles

L.G. Voskressensky (✉)
Russian Peoples' Friendship University, Miklukho-Maklaya Street, 6, 117198, Moscow, Russia
e-mail: lvoskressensky@sci.pfu.edu.ru

compounds, via the intermediate Strecker reaction. We presume that the formation of 3-aminoimidazoles can be explained by Scheme 1. The yields of obtained 3-aminoimidazoles were 15–79%.

References

Varma RS, Kumar D. (1999) Microwave-accelerated three-component condensation reaction on clay: solvent-free synthesis of imidazo[1,2-*a*] annulated pyridines, pyrazines and pyrimidines. Tetrahedron Lett 40:7665.

Ireland SM, Tye H, Whittaker M. (2003) Microwave-assisted multi-component synthesis of fused 3-aminoimidazoles. Tetrahedron Lett 44:4369.

Schwerkoske J, Perun T, Hulme C. (2005) New multi-component reaction accessing 3-aminoimidazo[1,2-*a*]pyridines. Tetrahedron Lett 46:8355.

Efficient Construction of (Aza)Indolyl Ynones by Pd/Cu-Catalyzed Multi-component Reactions

Eugen Merkul, Dieter Dorsch, and Thomas J. J. Müller

Indoles and 7-azaindoles bearing heterocyclic substituents in the 3-position constitute a very important class of biologically active small molecules and natural products, such as meridianins and variolins with their characteristic 2-amino pyrimidyl substitutents. For the construction of the pyrimidine rings, we envisioned their formation from the corresponding indolyl ynones. An efficient catalytic approach to these versatile intermediates required the development of two complementary multi-component approaches based upon the Sonogashira cross-coupling reaction. The carbonylative

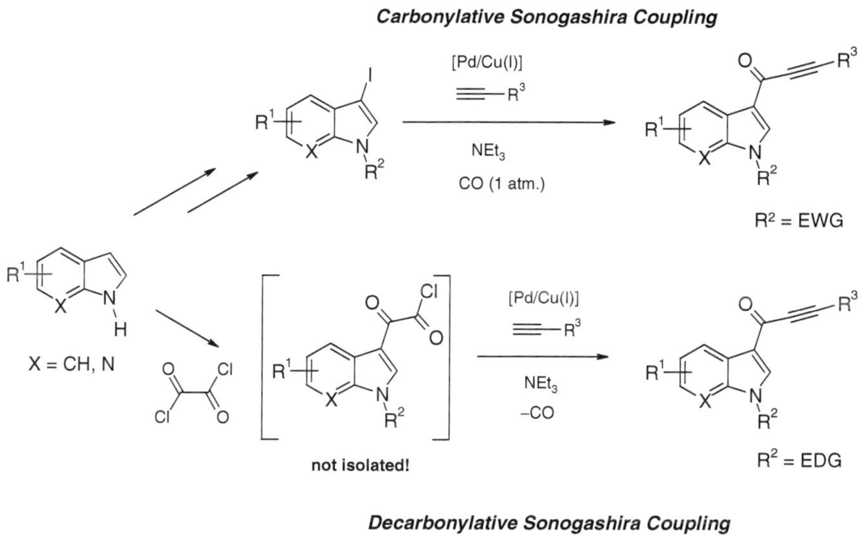

E. Merkul (✉)
Institut für Organische Chemie und Makromolekulare Chemie, Heinrich-Heine-Universität Düsseldorf, Universitätsstrasse, 1, 40225, Düsseldorf, Germany
e-mail: E.Merkul@uni-duesseldorf.de

alkynylation under one atmosphere of carbon monoxide and unprecedented decarbonylative Sonogashira coupling as well as their implementation to the synthesis of pyrimidines and representative biological data of the latter will be presented.

References

Karpov AS, Müller TJJ. (2003) New entry to a three-component pyrimidine synthesis by TMS–Ynones via Sonogashira coupling. Org Lett 5:3451.

Karpov AS, Merkul E, Rominger F, Müller TJJ. (2005) Concise syntheses of meridianins by carbonylative alkynylation and a four-component pyrimidine synthesis. Angew Chem Int Ed 44:6951.

Multi-component Synthesis of Pyridazine and Pyridazinoquinazoline Derivatives in the Presence of 12-Tungstophosphoric Acid

Hassan Sheibani, Zeinab Esfandiarpoor, and Soheila Behzadi

The chemistry and pharmacology of pyridazines and pyridazinoquinazoline have recently received considerable interest. This can be readily realized from the vast number of papers and patents dealing with the synthesis, chemistry, and biological activities of these compounds. In continuing our interest in the synthesis of heterocyclic compounds, in this paper we wish to report a one-pot synthesis of 6-acetyl-3-amino-2,5-diaryl-2,5-dihydro-4-pyridazincarbonitrile derivatives and substituted pyridazinoquinazolines which were prepared in a three-component reaction of arylhydrazones, malononitrile or ethyl cyanoacetate, and aldehydes at ambient temperature in the presence of 12-tungstophosphoric acid (PW) as catalyst and a catalytic amount of triethylamine.

The structure of the compounds synthesized was determined on the basis of their elemental analyses, mass spectrometry, ^{1}H and ^{13}C NMR and IR spectroscopic data.

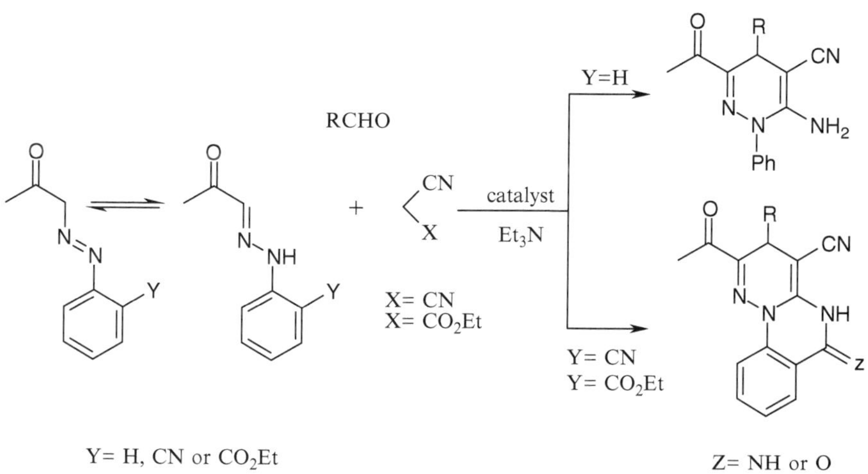

H. Sheibani (✉)
Department of Chemistry, Shahid Bahonar University of Kerman, Kerman 76169, Iran
e-mail: hsheibani@mail.uk.ac.ir

Reference

Nagawade RR, Khanna VV, Bhagwat SS, Shinde DB. (2005) Synthesis of new series of 1-aryl-1,4-dihydro-4-oxo-6-methyl pyridazine-3-carboxylic acid as potential antibacterial agents. Eur J Med Chem 40:1325.

Research of Nanosized Metal Oxide Action on Regio- and Stereoselectivity of the Multi-component Hantzsch Reaction

Julia A. Titova, Olga V. Fedorova, Irina G. Ovchinnikova, Marina S. Valova, Olga V. Koryakova, Gennady L. Rusinov, and Valery N. Charushin

Inorganic heterogeneous catalysis plays an important role in organic synthesis due to the facility of the recovery and reuse of solvent-insoluble catalysts. This property underlies the development of eco-friendly and economic processes.

For the first time, the influence of heterogeneous catalysis action on the regio- and stereoselectivity of the multicomponent Hantzsch reaction was investigated. In heterogeneous catalysis conditions, a one-pot Hantzsch reaction was shown to afford nitrendipine **4b** in a 70% yield, compared with a 10% yield without nanosized metal oxide. Further, it has been found that in heterogeneous catalysis conditions the reaction runs via the addition of enamine **1** to chalcone **2**, which is not fixed in conventional conditions.

The simultaneous use of metal nanooxide and chiral modifier results in an increase in the regio- and stereoselectivity of the Hantzsch reaction. Namely, the use of nanosized Al oxide in the Hantzsch one-pot synthesis results in nifedipine **4a** in an 87% yield. Also, a two-stage procedure for the synthesis of racemic nitrendipine **4b** in a 75% yield has been elaborated. New catalytic systems suggested allow one to change a ratio between stereoisomers in favor of the *S*-enantiomer with up to a 37% excess.

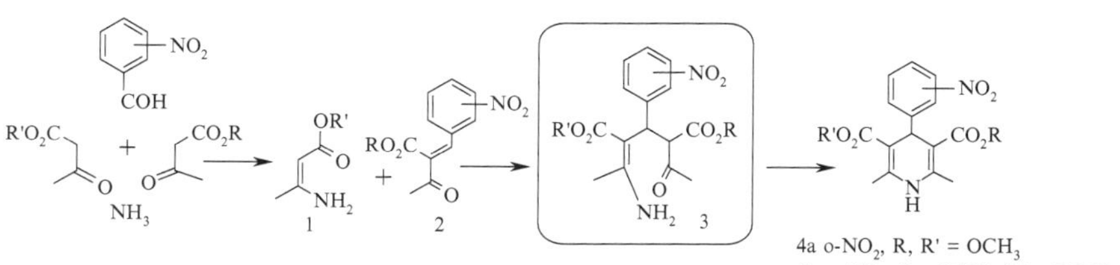

4a o-NO_2, R, R' = OCH_3
4b *m*-NO_2, R = OCH_3, R' = OC_2H_5

J.A. Titova (✉)
I.Ya. Postovsky Institute of Organic Synthesis of Russian Acaemy of Sciences, Urals Branch, S. Kovalevskoy/Akademicheskay Street, 22/20, 620041, Ekaterinburg, Russia
e-mail: titova@ios.uran.ru

Acknowledgment Our work has been executed with financial support from the Russian Foundation for Basic Research (grant nos. 07-03-96111 and 07-03-96113); the Presidium of the Russian Academy of Science (under the project "Heterogeneous Catalysts with the Use of Nanosized Metals and Their Oxides"), and also the Korean Atomic Energy Research Institute (contract 01/06, "Development of New Chiral Catalytic Systems on the Basis of Nanocrystalline Metals and Metal Oxides").

References

Fedorova OV, Ovchinnikova IG, Kharchuk VG, Rusinov GL, Uimin MA, Mysik AA, Yermakov AY, Uhm YR, Rhee CK, Charushin VN. (2007) Investigation of the nanosized metal oxides as stereoselective catalysts of the Hantzsch and Biginelli reactions. Abstracts of the 8th International Symposium on Nanocomposites and Nanoporous Materials, Jeju Haevichi Resort, Korea, pp. 181–182.

Titova JA, Fedorova OV, Ovchinnikova IG, Valova MS, Koryakova OV, Rusinov GL, Charushin VN. (2009) Research of the mechanism of nanosized metal oxide action on regio- and stereo-selectivity of the Hantzsch reaction. Abstracts of the VIII International Conference on Mechanisms of Catalytic Reactions, Novosibirsk, pp. I–95.

Short Communication: Synthesis

Nanocrystalline Copper (II) Oxide-Catalyzed One-pot Synthesis of Polyfunctionalized Pyrans via a Three-Component Cyclocondensation Under Solvent-Free Conditions

Rahim Hekmatshoar and Sodeh Sadjadi

The 4*H*-pyran group is a constituent of the structures of a series of natural products with interesting biological and pharmacological activities such as anticoagulant, anticancer, spasmolytic, and antianaphylactic properties. Furthermore, these compounds can be employed as pigments and photoactive materials and utilized as potential biodegradable agrochemicals.

Nanocrystalline copper oxide catalyzes the facile condensation of diketo esters, aldehydes, and malononitrile under mild conditions to afford the corresponding 2-amino-4*H*-pyran derivatives in good to high yields. The catalyst can be prepared easily, is stable and storable, is easily recycled, and can be reused for several cycles with consistent activity.

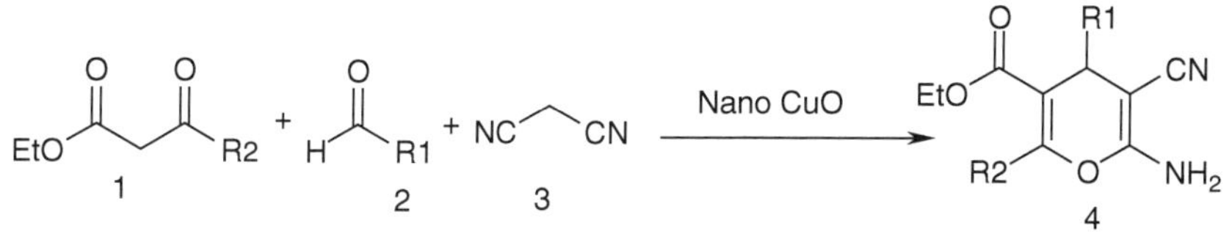

R. Hekmatshoar (✉)
Department of Chemistry, School of Science, Alzahra University, Vanak, Tehran, Iran
e-mail: rhekmatus@yahoo.com

Session 2
Multi-Component Reactions in Heterocyclic Chemistry

Multi-component Synthesis of Heterocycles by Palladium-catalyzed Generation of Alkynones, Alkenones and Allenes

Thomas J.J. Müller

Multi-component and domino reactions are efficient and effective methods in the sustainable and diversity-oriented synthesis of heterocycles. In particular, transition metal-catalyzed multi-component sequences have recently gained considerable interest. Based upon the Sonogashira entry to alkynones, alkenones, and intermediate allenes, we have opened new avenues to the one-pot synthesis of numerous classes of heterocyclic frameworks in an MCR fashion. This methodological approach has now found various applications in one-pot syntheses of functional chromophores, pharmaceutically active compounds, and marine alkaloids and derivatives.

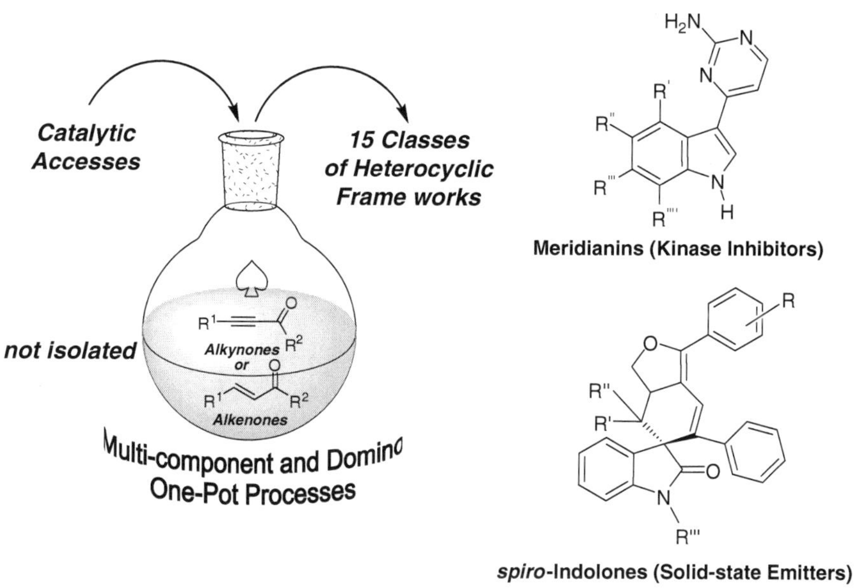

T.J.J. Müller (✉)
Lehrstuhl für Organische Chemie, Institut für Organische Chemie und Makromolekulare Chemie, Heinrich-Heine-Universität Düsseldorf, Universitätsstrasse 1, D-40225, Düsseldorf, Germany
e-mail: ThomasJJ.Mueller@uni-duesseldorf.de

M.A. Mironov (ed.), *MCR 2009*, Advances in Experimental Medicine and Biology 699, DOI 10.1007/978-1-4419-7270-5_2, © Springer Science+Business Media, LLC 2011

References

D'Souza DM, Müller TJJ. (2007) Multi-component syntheses of heterocycles by transition-metal catalysis. Chem Soc Rev 36:1095–1120.

Willy B, Müller TJJ. (2008) Consecutive multi-component syntheses of heterocycles via palladium–copper catalyzed generation of alkynones. ARKIVOC Part I:195–212.

Schramm OG, Oeser T, Kaiser M, Brun R, Müller TJJ. (2008) Rapid one-pot synthesis of anti-parasitic quinolines based upon the microwave-assisted coupling-isomerization reaction (MACIR). Synlett 2008:359–362.

Multi-component Reactions as Useful Platforms to Explore the Chemical Space

Romano V. A. Orru

The rapid generation of diverse sets of complex molecules can be achieved by employing diversity-oriented synthetic strategies in combination with so-called complexity-generating reactions. Multi-component reactions (MCRs), which combine in one pot at least three simple building blocks, provide a most powerful platform to access diversity as well as complexity in a limited number of reaction steps. Here we describe novel modular reaction sequences based on our previously reported MCR chemistry in combination with other common organic reactions or even with a second MCR. The combination of our MCRs with, e.g., cycloadditions, transition-metal mediated cross-coupling reactions, or more traditional MCRs, such

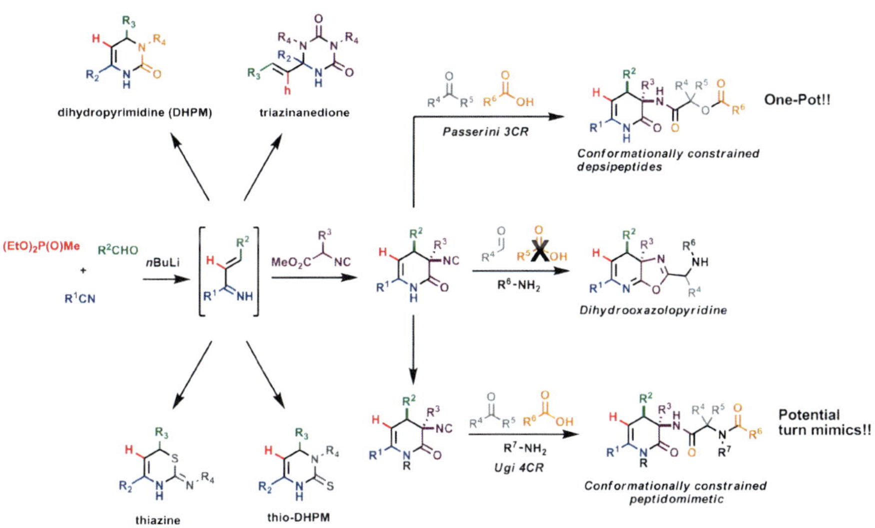

R.V.A. Orru (✉)
Department of Chemistry and Pharmaceutical Sciences, VU University, Amsterdam, The Netherlands
e-mail: orru@few.vu.nl

as the Biginelli-3CR, the Ugi-4CR, and the Passerini-3CR, will be shown. Examples in which the synthetic methodology was applied for the easy generation of focused libraries for the synthesis of medicinally relevant ligands will be discussed.

References

Bon RS, Vliet van B, Sprenkels NE, Schmitz RF, Kanter de FJJ, Stevens CV, Swart M, Bickelhaupt FM, Groen MB, Orru RVA. (2005) Multicomponent synthesis of 2-imidazolines. J Org Chem 70:3542–3546.

Orru RVA, de Greef M. (2003) Recent advances in solution-phase multicomponent methodology for the synthesis of heterocyclic compounds. Synthesis 10:1471–1499.

Paravidino M, Bon RS, Scheffelaar R, Vugts DJ, Znabet A, Kanter de FJJ, Lutz M, Spek AL, Groen MB, Orru RVA. (2006) Diastereoselective multicomponent synthesis of dihydropyridones with an isocyanide functionality. Org Lett 8:5369–5372.

Elders N, Schmitz RF, de Kanter FJJ, Ruijter E, Groen MB, Orru RVA. (2007) A resource-efficient and highly flexible procedure for a three-component synthesis of 2-imidazolines. J Org Chem 72:6135–6142.

Groenendaal B, Vugts DJ, Schmitz RF, de Kanter FJJ, Ruijter E, Groen MB, Orru RVA. (2008) A multicomponent synthesis of triazinane diones. J Org Chem 73:719–725.

Groenendaal B, Ruijter E, Orru RVA. (2008) 1-Azadienes in cycloaddition and multicomponent reactions towards N-heterocycles. Chem Commun (43):5474–5476.

Multi-component Heterocyclizations: Control of Chemo- and Regioselectivity

Valentin A. Chebanov, Yana I. Sakhno, Vyacheslav E. Saraev, Elena A. Muravyova, Anastasia Yu. Andrushchenko, and Sergey M. Desenko

The control of selectivity, for example chemo- and regioselectivity, is among the most important objectives in organic chemistry. For multi-component reactions involving the simultaneous molecular interaction of three or more components, the issue of selectivity is of particular significance due to the high probability of several potential parallel reaction pathways leading to different product classes. Many different process parameters such as temperature, pressure, solvent, catalyst type, microwave and ultrasonic irradiations, and other factors can be utilized to modulate the selectivity of synthetic transformations.

In the present report some of our results in the tuning of chemo- and regioselectivity of multi-component reactions involving active methylene compounds, carbonyls, and aminoazoles, containing several nonequivalent nucleophilic reaction centers, are observed. The general principles of the direction control for this type of interaction are reported, and synthetic methodology allowing high selectivity in obtaining certain types of heterocycles is presented.

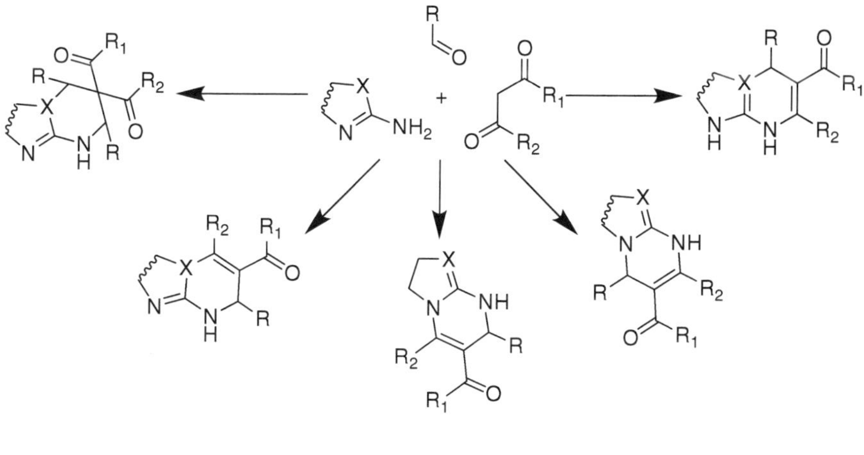

V.A. Chebanov (✉)
Department of Heterocyclic Compounds Chemistry, Division of Functional Materials Chemistry, SSI "Institute for Single Crystals" NAS of Ukraine, Lenin Avenue, 60, Kharkiv, 61001, Ukraine
e-mail: chebanov@isc.kharkov.com

References

Chebanov VA, Sakhno YI, Desenko SM, Shishkina SV, Musatov VI, Shishkin OV, Knyazeva IV. (2005) Three-component procedure for the synthesis of 5-aryl-5,8-dihydroazolo[1,5-a]pyrimidine-7-carboxylic acids. Synthesis (15):2597.

Chebanov VA, Muravyova EA, Desenko SM, Musatov VI, Knyazeva IV, Shishkina SV, Shishkin OV, Kappe CO. (2006) Microwave-assisted three-component synthesis of 7-aryl-2-alkylthio-4,7-dihydro-1,2,4-triazolo[1,5-*a*]-pyrimidine-6-carboxamides and their selective reduction. J Comb Chem 8:427.

Chebanov VA, Saraev VE, Desenko SM, Chernenko VN, Knyazeva IV, Groth U, Glasnov T, Kappe CO. (2008) Tuning of chemo- and regioselectivities in multicomponent condensations of 5-aminopyrazoles, dimedone, and aldehydes. J Org Chem 73:5110.

Sakhno YI, Desenko SM, Shishkina SV, Shishkin OV, Sysoyev DO, Groth U, Kappe CO, Chebanov VA. (2008) Multicomponent cyclocondensation reactions of aminoazoles, arylpyruvic acids and aldehydes with controlled chemoselectivity. Tetrahedron 64:11041.

Multi-component Reactions of H_2S with Aldehydes and Amines as an Efficient Route to Heterocycles and Thioaza-Crown Compounds

V. R. Akhmetova, G. R. Khabibullina, E. B. Rakhimova, R. A. Vagapov, R.R. Khairullina, Z.T. Niatshina, and N.N. Murzakova

The approaches to the synthesis of sulfur- and nitrogen-containing mono- and poly-heterocycles obtained by condensation of amines and hydrazines with thiomethylating reagent "CH_2O–H_2S".

Nowadays our attention focuses on the investigation of synthetic possibilities of multi-component reactions of H_2S with different carbonyl compounds, amines, and hydrazines to produce new types of heterocyclic compounds, such as substituted thiadiazinanes (**1**), dithiadiazacyclooctanes (**2**), and thiadiaza- (**3**) and dithiadiazabicyclancs (**4**).

In this report the results of investigations devoted to the *one-pot* design of thioaza-crown compounds are considered. The rules which define intermolecular cyclocondensation of binucleophilic starting compounds with the "CH_2O–H_2S" reagent will be discussed.

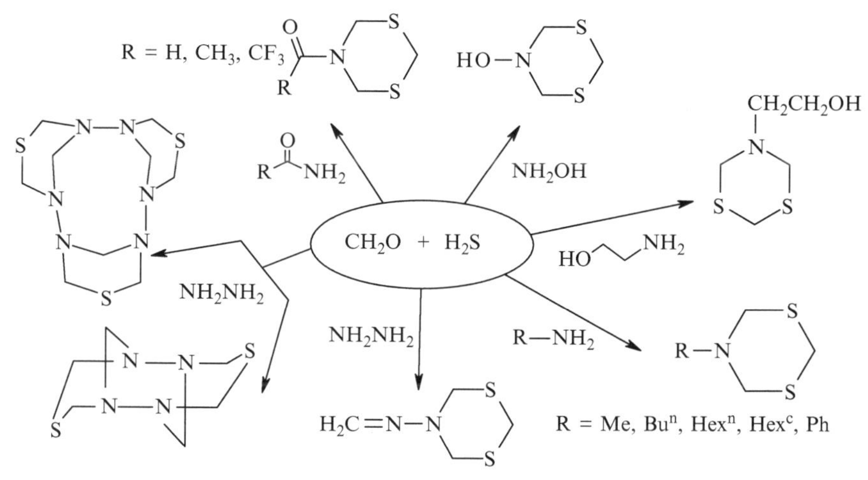

V.R. Akhmetova (✉)
Institute of Petrochemistry and Catalysis of Russian Academy of Sciences, 450075 Ufa, prosp. Oktyarya 141, Russia
e-mail: ink@anrb.ru

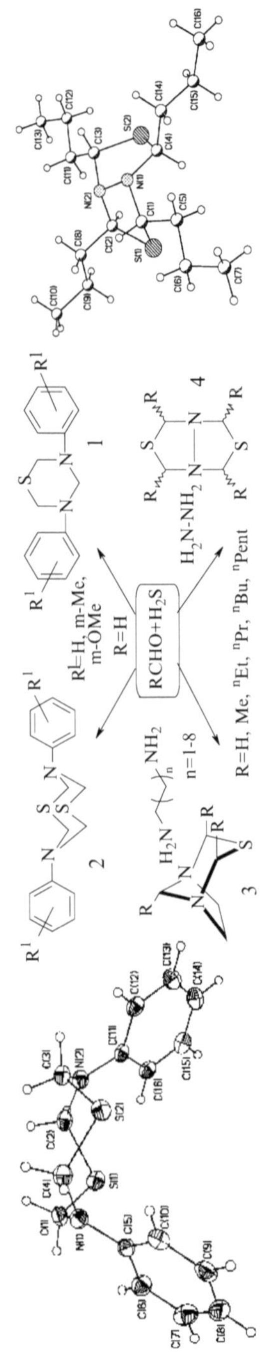
RCHO+H2S
R=H
R1=H, m-Me, m-OMe
1
2
3
4
H2N-NH2
n=1-8
R=H, Me, nEt, nPr, nBu, nPent

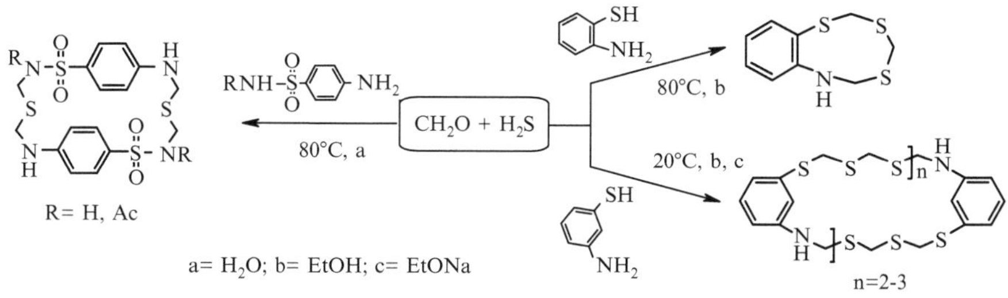

Sulfur- and nitrogen-containing heterocycles are of special interest as potential antibacterial and antivirus preparations and also as selective sorbents and flotation agents for metals and complexones in supramolecular chemistry.

References

Akhmetova VR, et al. (2004) Multicomponent heterocyclization of hydrazine, hydrogen sulfide, and formaldehyde. Russ Chem Bull Int Ed 8:1717.

Akhmetova VR, et al. (2004) Multicomponent condensation of aliphatic amines with formaldehyde and hydrogen sulfide. Russ Chem Bull Int Ed 2:432.

Akhmetova VR, et al. (2007) Thiomethylation of amino alcohols using formaldehyde and hydrogen sulfide. Russ J Org Chem 6:919.

Cascade Recyclizations of 1*H*-Pyrrole-2, 3-diones: A Method of Unusual Heterocyclic Systems Construction

Andrey N. Maslivets

Monocyclic 1*H*-pyrrole-2,3-diones and 1*H*-pyrrole-2,3-diones, in which one pyrrolidone cycle is annexed to an azaheterocyclic fragment, under the action of NH, OH, SH, CH *bi*-nucleophilic reagents undergo cascade recyclizations with unusual condensed, bridged, and spiro-*bis*-azaheterocyclic systems formation. Some examples of the formed compounds that have resulted appear below.

Cycloaddition reactions of 1*H*-pyrrole-2,3-diones and heterocumulenes, generated on their basis, represent a convenient way of unusual condensed azaheterocyclic systems and ensembles of heterocyclic systems formation. Below are some examples of the formed compounds that have resulted.

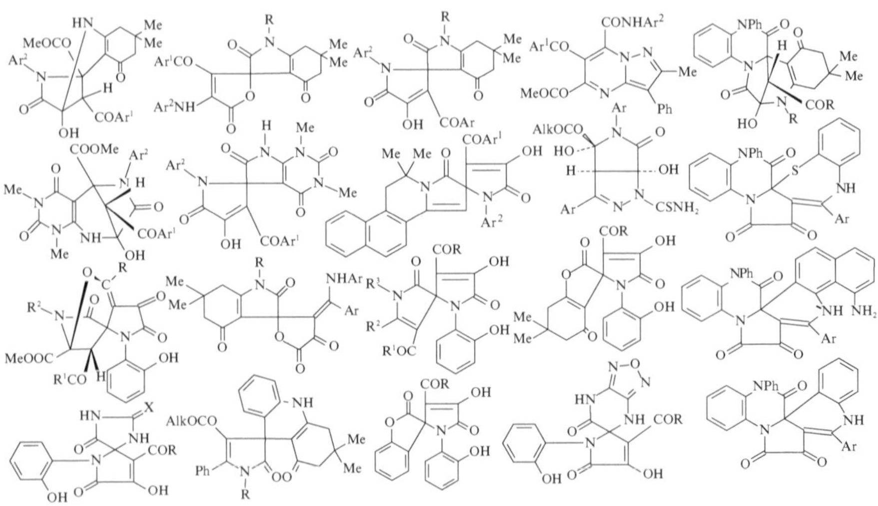

A.N. Maslivets (✉)
Perm State University, Bukirev Street, 15, 614990, Perm, Russia
e-mail: koh2@psu.ru

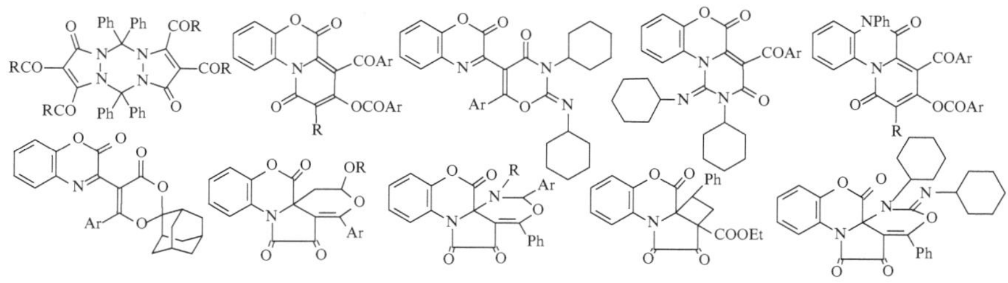

In the report, schemes of the mentioned transformations will be presented.

Acknowledgment The present work was supported by the Russian Foundation for Basic Research (grant nos. 07-03-96036 and 08-03-01032).

References

Silaichev PS, Dmitriev MV, Aliev ZG, Maslivets AN. (2010) Five-membered 2,3-dioxohetero-cycles: LXVII. Pyrroledione-pyrroledion recyclization of isopropyl 2-(1-aryl-4,5-dioxo-2-phenyl-4,5-dihydro-1H-pyrrol-3-yl)-2-oxoacetates under the action of arylamines. Crystal and molecular structure of (z)-isopropyl 2-hydroxy-4,5-dioxo-1-phenyl-3-[phenyl(phenylamino)-methylene]pyrrolidine-2-carboxylate. Russ J Org Chem 46(2):255–259.

Denislamova ES, Maslivets AN. (2010) Five-membered 2,3-dioxoheterocycles: LXVIII. Three pathways in the reaction of methyl 3-aroyl-1-aryl-4,5-dioxo-4,5-dihydro-1H-pyrrole-2-car-boxylates with 3-amino-5,5-dimethylcyclohex-2-en-1-one. Russ J Org Chem 46(3):389–393.

Khalturina VV, Shklyaev YS, Aliev ZG, Maslivets AN. (2010) Five-membered 2,3-dioxohetero-cycles: LXIX. Direct heterocyclization of [3,4-dihydroisoquinolin-1(2*H*)-ylidene]-acetamides with 5-arylfuran-2,3-diones. Crystalline and molecular structure of (3E,5Z)-3-[3,3-dimethyl-3,4-dihydroisoquinolin-1(2*H*)-ylidene]-5-(2-oxo-2-phenylethylidene)pyrrolidine-2,4-dione. Russ J Org Chem 46(4):539–542.

Alkyne-Induced Tandem Transformations of (Hetero)Annulated Five-, Six-, or Seven-Membered *N*-Heterocycles and a New MCR on This Motif

Leonid G. Voskressensky

Medium-sized N-containing heterocycles, in particular eight- and nine-membered rings, are key structures of various structurally remarkable natural products. As the direct formation of these ring sizes from acyclic precursors is entropically and enthalpically disfavored, the efficient construction of medium-sized cycles is a challenge and has therefore attracted considerable attention in recent years. We have recently reported tetrahydropyridine (THPy) and tetrahydroazepine ring expansion in tetrahydropyrrolopyridines under the action of activated alkynes. Applying this methodology to other THP and tetrahydroazepine-containing substrates showed that this reaction is general for a number of heterocyclic systems (Scheme 1).

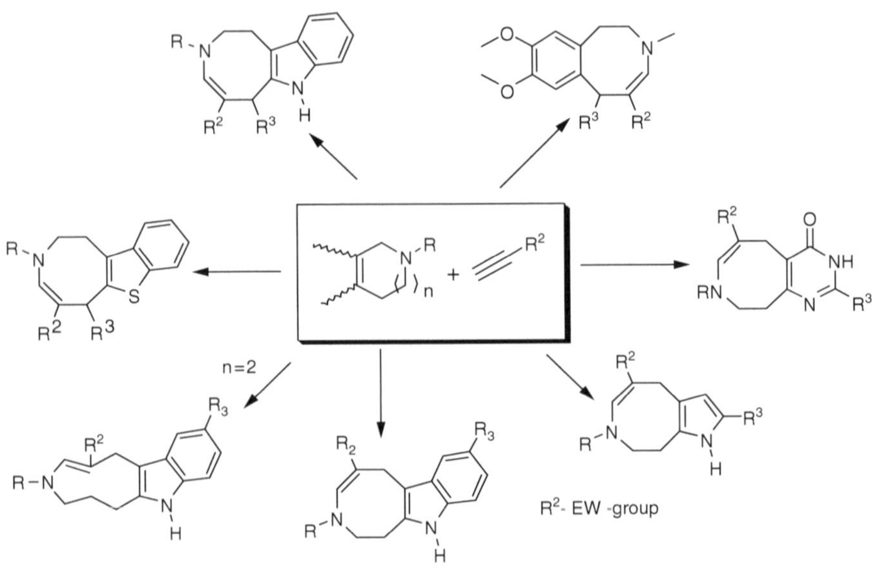

Scheme 1 Synthesis of medium-size heterocycles

L.G. Voskressensky (✉)
Russian Peoples' Friendship University, Miklukho-Maklaya Street, 6, 117198, Moscow, Russia
e-mail: lvoskressensky@sci.pfu.edu.ru

We have elaborated a new isocyanide-based MCR representing the combination of the Passerini reaction with the above-described tetrahydropyridine ring enlargement process (Scheme 2). The scope and limitations of this MCR will be discussed.

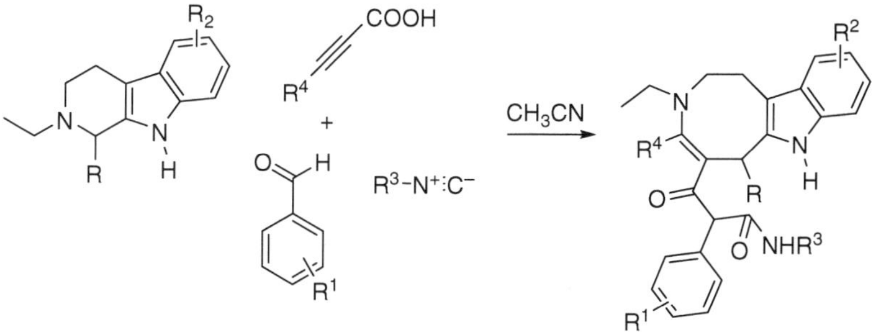

Scheme 2 Tandem of the Passerini reaction and tetrahydropyridine ring enlargement

References

Voskressensky LG, Borisova TN, Kulikova LN, Varlamov AV, Catto M, Altomare C, Carotti A. (2004) Tandem cleavage of hydrogenated β- and γ-carbolines – new practical synthesis of tetrahydroazocino[4,5-*b*]indoles and tetrahydroazocino[5,4-*b*]indoles showing acetylcholinesterase inhibitory activity. Eur J Org Chem 2004(14):3128–3135.

Voskressensky LG, Kulikova LN, Borisova TN, Varlamov AV. (2008) Synthesis of heteroannulated azocine derivatives. Adv Heterocycl Chem 96:81–122.

N–C and *C–C* Recyclizations of Pyrimidinium Salts

Gevorg G. Danagulyan, Armen D. Murtchyan, and Araksya K. Tumanyan

The urgency of the study of pyrimidine systems' nucleophilic rearrangements is doubtless. They attract researchers' attention by the fact that they are not always a priori evident and frequently unpredictable. Accompanied with the complex reconstruction of a molecule skeleton resulting in a cardinal change of configuration, they are nevertheless one-stage processes from the preparative point of view that just predetermines the interest toward the study of similar transformations. Recent publications have shown that these reactions have turned into an important method of organic synthesis and have been applied for the transformation of various biologically active and medicinal compounds – nucleosides and drugs.

We have studied in detail rearrangements of recyclizational transformations of pyrimidines into pyridine derivatives (the so-called Kost–Sagitullin rearrangements). In these reactions the nitrogen atom of a pyrimidine ring is substituted by the exocyclic carbon atom being in the second position (*N–C* substitution or *N–C* recyclizations) (Scheme 1); another type of nucleophilic recyclization of pyrimidines,

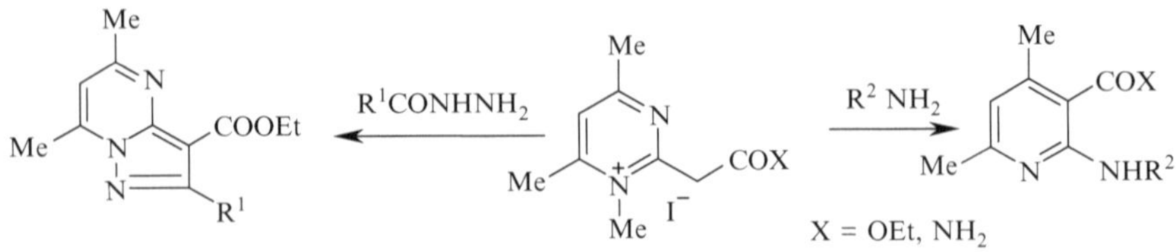

Scheme 1 Kost-Sagitullin rearrangement

G.G. Danagulyan (✉)
Russian-Armenian (Slavonic) State University, 123 H. Emin Street, 375051, Yerevan, Republic of Armenia
and
Institute of Organic Chemistry, NAS RA, Yerevan, Armenia
e-mail: gdang@email.com

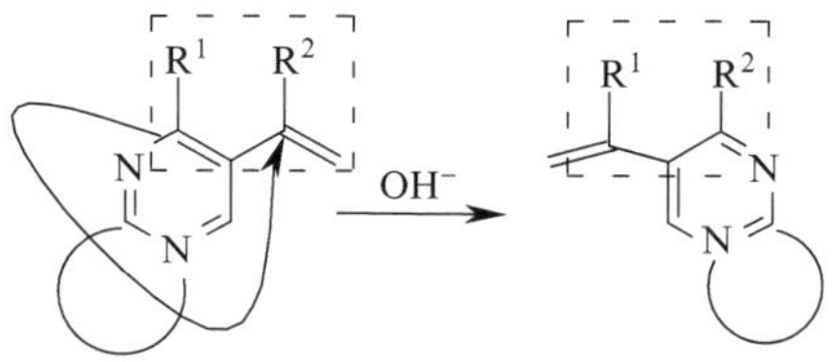

Scheme 2 *C–C* recyclization

which occurs with a heterocycle substitution of the carbon atom by the exocyclic carbon atom, is the *C–C* recyclization (Scheme 2).

In developing our investigations of nucleophilic recyclizations, we studied transformations of pyrimidines on the models that are potentially able to competitively undergo various types of recyclizations – both Dimroth rearrangements and/or Kost–Sagitullin rearrangements, and *C–C* recyclizations. As a result of the reaction of pyrimidinium salt **1** with alcoholic methylamine, we managed to isolate from the reaction mixture a compound in 55% yield methylimine of 2-amino-4-hydroxy-6-methylamino-5-phenylpyridine-3-aldehyde (**2**). The same salt **1**, when heated in benzylamine, transformed into a mixture of three compounds: the product of destructive aminolysis (33%) and two products of C–C recyclization, pyrimidines **3** and **4** (Scheme 3).

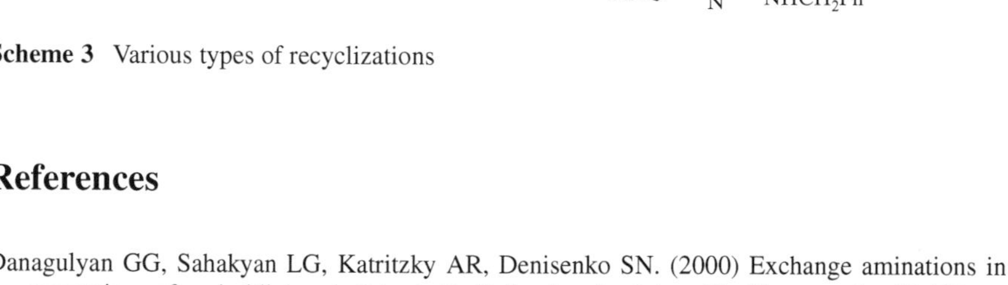

Scheme 3 Various types of recyclizations

References

Danagulyan GG, Sahakyan LG, Katritzky AR, Denisenko SN. (2000) Exchange aminations in conversions of pyrimidinium iodides to 2-alkylaminonicotinic acids. Heterocycles 53:419.

Danagulyan GG. (2005) Kost–sagitullin rearrangement and other isomerization recyclizations of pyrimidines. (Review). Chem Heterocycl Comp 41:1205.

Synthetic Approach to Highly Functionalized Mesocyclic Heterocycles by Coupling an Ugi or Passerini Reaction (PADAM Strategy) with a Pd-Mediated Cyclocarbonylation

Luca Banfi, Andrea Basso, Fabio De Moliner, Giuseppe Guanti, Elena Petricci, Renata Riva, and Maurizio Taddei

Multi-component reactions are useful tools for the synthesis of innovative and very complex scaffolds according to a convergent approach, providing a molecule that contains fragments derived from all the building blocks employed in the MCR condensation in a single synthetic step. As they can be followed by postcondensation transformations exploiting additional functional groups present in the building blocks and inert during the MCR step, they are particularly suitable for diversity-oriented synthesis.

In this context, we explored the possibility of submitting MCR-derived precursors, endowed with a bromine on an aryl moiety, to an intramolecular cyclocarbonylation reaction, in order to develop a fast new entry to highly functionalized and unusual N-heterocycles.

For this purpose we synthesized a series of compounds, equipped with a free alcoholic hydroxy group or a free secondary amino group. This task was accomplished by exploiting either the PADAM (Passerini-amine deprotection-acyl migration) strategy or an Ugi reaction followed by an unusual rearrangement that leaves the amine-derived nitrogen free to undergo the following process. In both cases we also placed the required bromine atom either on the isocyanide or on the amine moiety.

The acyclic precursors were then submitted to the cyclocarbonylation reaction promoted by a Pd catalyst under carbon monoxide atmosphere and microwave irradiation. The results of these preliminary tests will be discussed.

F. De Moliner (✉)
Università di Genova, Dipartimento di Chimica e Chimica Industriale,
Via Dodecaneso 31, 16146, Genova, Italy
e-mail: fabio.demoliner@libero.it

References

Burke MD, Schreiber SL. (2004) A planning strategy for diversity-oriented synthesis. Angew Chem Int Ed 43:46–58.

Banfi L, Basso A, Guanti G, Riva R. (2003) Passerini reaction – amine deprotection – acyl migration (PADAM): a convenient strategy for the solid-phase preparation of peptidomimetic compounds. Mol Div 6:227–235 and articles quoted within.

Zhu J, Bienaymé H. (2005) Multicomponent reactions. Weinheim: Wiley.

Solvent-Free Three-Component Condensation of Aldehydes and Malononitrile with Naphthols Affording 2-Amino-4*H*-chromenes

M. Reza Naimi-Jamal, Sara Mashkouri, and Ali Sharifi

Solvent-free and one-pot multi-component condensations represent very powerful green chemical technology procedures from both economical and synthetic points of view and represent a possible instrument to perform a near-ideal synthesis.

2-Amino-4*H*-chromenes represent an important class of compounds; they are the main components of many naturally occurring products and are generally prepared by refluxing malononitrile, an aldehyde, and an activated phenol in the presence of hazardous organic bases such as piperidine in organic solvents such as ethanol and acetonitrile for several hours.

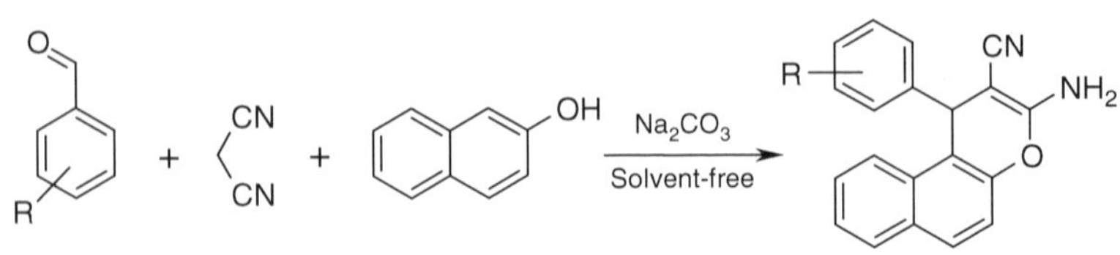

Herein we wish to report the three-component condensation of aldehydes and malononitrile with α- and β-naphthols with excellent yields.

General Procedure for 2-Amino-4*H*-Chromenes

In a typical experiment, a stoichiometric mixture of an aldehyde (**1**), malononitrile (**2**), and β-naphthol (**3**) (1.0 mmol each) and sodium carbonate (0.1 mmol) were mixed using a mortar and pestle. The resulting mixture was heated in a drying oven at

M.R. Naimi-Jamal (✉)
Organic Chemistry Research Laboratory, Department of Chemistry, Iran University of Science and Technology, 16846, Tehran, Iran
e-mail: naimi@iust.ac.ir

125°C. After cooling, the mixture was washed with hot water and purified by recrystallization from hot ethanol, if necessary.

Different 2-amino-4*H*-chromenes were prepared with a similar method. The procedure is very simple, efficient, and environmentally friendly, as it does not use any solvent and toxic catalyst.

References

Kaupp G. (2005) Organic solid-state reactions with 100% yield. Top Curr Chem 254:95–183.

Kemnitzer W, Kasibhatla S, Jiang S, Zhang H, Zhao J, Jia S, Xu L, Crogan-Grundy C, Denis R, Barriault N, Vaillancourt L, Charron S, Dodd J, Attardo G, Labrecque D, Lamothe S, Gourdeau H, Tseng B, Drewe J, Cai SX. (2005) Discovery of 4-aryl-4H-chromenes as a new series of apoptosis inducers using a cell- and caspase-based high-throughput screening assay. 2. Structure–activity relationships of the 7- and 5-, 6-, 8-positions. Bioorg Med Chem Lett 15:4745–4751.

Anderson DR, Hegde S, Reinhard E, Gomez L, Vernier WF, Lee L, Liu S, Sambandam A, Snider PA, Masih L. (2005) Aminocyanopyridine inhibitors of mitogen activated protein kinase-activated protein kinase 2 (MK-2). Bioorg Med Chem Lett 15:1587–1590.

New Pyrido[3,2-*c*]pyridones and Pyrido[3,2-*c*] pyrazoles Accessible by a One-Step Multi-component Synthesis

Nikolai M. Przhevalski, Elena N. Rozhkova, Gennadii P. Tokmakov, and Igor V. Magedov

We have recently initiated a research program aimed at the structural simplification of natural products, specifically by utilizing multi-component synthetic processes. We showed that the stereochemically complex structure of an important anticancer lead podofillotoxin can be efficiently simplified to a dihydropyridopyrazole scaffold, which is accessible via a one-step multi-component synthetic reaction. The resulting library of compounds retains a significant portion of podofillotoxin's

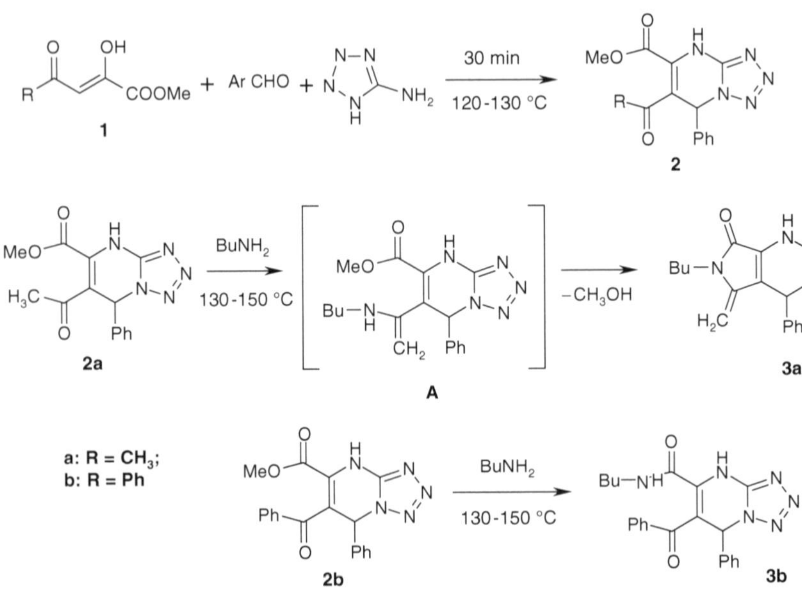

N.M. Przhevalski (✉)
Russian State Agrarian University, Moscow Timiryazev Agricultural Academy, Timiryazevskaya ul., 49, 127550, Moscow, Russia
e-mail: prjevalski@mail.ru

cytotoxic potency and apoptosis-inducing potential. In these efforts, we have been investigating a compound library based on pyrano[3,2-*c*]pyridine (**1**, *X*=O) and pyrano[3,2-*c*]quinolone scaffolds. A three-component reaction of pyridon (or quinolone) with malononitril and various aromatic aldehydes in a 1:1:1 ratio proceeds smoothly in refluxing ethanol containing a small quantity of Et_3N.

To develop compound libraries, we considered a reaction of 4-hydroxypyridones (as well as pyrazolones) with aldehydes, malononitril, and ammonium acetate. Indeed, we found that the desired compounds, pyrido[3,2-*c*]pyridones (**2**) and pyridopyrazoles (**3**), can be synthesized via a four-component reaction.

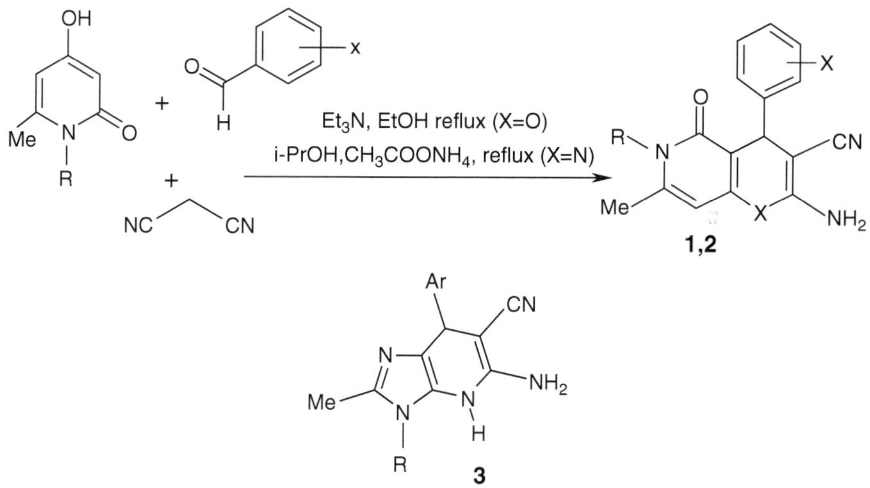

References

Magedov IV, Manpadi M, Rozhkova EN, Przheval'skii NM, Rogelj S, Shors ST, Steelant WFA, Van Slambrouck S, Kornienko A. (2007) Structural simplification of bioactive natural products with multicomponent synthesis: dihydropyridopyrazole analogues of podophyllotoxin. Bioorg Med Chem Lett 17:1381.

Magedov I, et al. (2008) Structural simplification of bioactive natural products with multicomponent synthesis. 2. Antiproliferative and antitubulin activities of pyrano[3,2-c]pyridones and pyrano[3,2-c]quinolones. J Med Chem 51:2561.

Magedov IV, Manpadi M, Van Slambrouck S, Steelant WFA, Rozhkova E, Przheval'skii NM, Rogelj S, Kornienko A. (2007) Discovery and investigation of antiproliferative and apoptosis-inducing properties of new heterocyclic podophyllotoxin analogues accessible by a one-step multicomponent synthesis. J Med Chem 50:5183.

Synthesis of 1-Hydroxyaryl-4-Acyl-5-Aryl-3-Hydroxy-3-Pyrrolin-2-Ones

M. N. Armisheva, N. A. Rassudihina, M. I. Vahrin, and V. L. Gein

Multi-component reactions have attracted enormous interest. Among the multi-component reactions, the three-component processes have been developed into useful organic procedures. The 3-hydroxy-3-pyrrolin-2-one derivatives are important due to their therapeutic and pharmacological properties.

Our investigations have shown that equivalent amounts of methylic esters of acylpyruvic acids, 4-aminophenoles, and aromatic aldehyde in glacial acid, with good yields, lead to 1-hydroxyaryl-4-acyl-5-aryl-3-hydroxy-3-pyrrolin-2-ones (**I–XXX**).

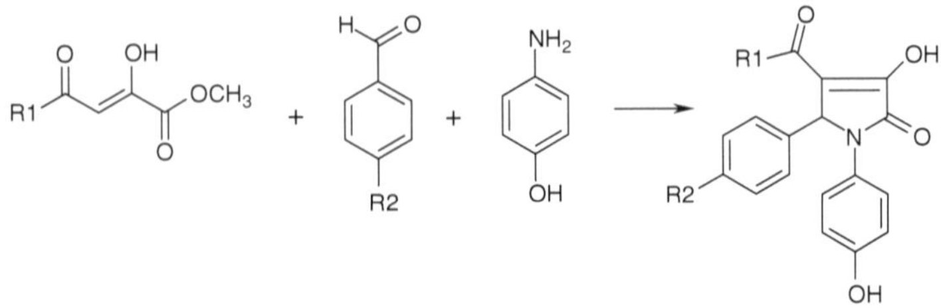

I-XXX

I (R^1 = Ph, R^2 = H), **II** (R^1 = H, R^2 = H), **III** (R^1 = H, R^2 = 4-Cl), **IV** (R^1 = Ph, R^2 = 4-Cl), **V** (R^1 = H, R^2 = 2-NO_2), **VI** (R^1 = H, R^2 = 2-OCH_3), **VII** (R^1 = H, R^2 = 4-CH_3), **VIII** (R^1 = H, R^2 = 2-NO_2), **IX** (R^1 = H, R^2 = 2-Cl), **X** (R^1 = H, R^2 = 2,5(OCH_3)$_2$), **XI** (R^1 = H, R^2 = 2-F), **XII** (R^1 = H, R^2 = 4-C_3H_7), **XIII** (R^1 = H, R^2 = 4-NO_2), **XIV** (R^1 = H, R^2 = 3-OCH_3), **XV** (R^1 = H, R^2 = Ph), **XVI** (R^1 = H, R^2 = 4-F), **XVII** (R^1 = H, R^2 = 3-OCH_3-4-OH), **XVIII** (R^1 = H, R^2 = 3-F), **XIX** (R^1 = Ph, R^2 = 3,4-(OCH_3)$_2$), **XX** (R^1 = Ph, R^2 = 2-NO_2), **XXI** (R^1 = Ph, R^2 = 3-NO_3), **XXII** (R^1 = 4-CH_3, R^2 = 4-CH_3), **XXIII** (R^1 = 4-CH_3, R^2 = Ph), **XXIV** (R^1 = 4-CH_3, R^2 = 3-F), **XXV** (R^1 = 4-CH_3, R^2 = 2-Cl), **XXVI** (R^1 = 4-CH_3, R^2 = 4-NO_2), **XXVII** (R^1 = Ph, R^2 = Ph), **XXVIII** (R^1 = 4-CH_3, R^2 = 4-F), **XXIX** (R^1 = 4-CH_3, R^2 = 4-Br), **XXX** (R^1 = 4-CH_3, R^2 = 4-OCH_3).

N.A. Rassudihina (✉)
P erm State Pharmaceutical Academy, 614990 Lenin Street, 48 Perm, Russia
e-mail: limans@list.ru

The proposed structures of the synthesized compounds were confirmed by IR and ^{1}H NMR spectroscopy.

Reference

Gein VL. (2004) Tetraghydropyrrol- and tetraghydrofuran-2,3-diones. Perm, Russia: PSPA.

Synthesis of Highly Stable, Unusual Charge-Separated Pyridinium-, Isoquinolinium-, Quinolinium-, and N-Methylimidazolium-Tetronic Acid Zwitterions

Ahmad Shaabani, Ali Hossein Rezayan, Afshin Sarvary, Marjan Heidary, and Seik Weng Ng

In connection of our current studies on multi-component reactions involving zwitterionic species, and of our interest in the chemistry of tetronic acid, herein we would like to present a unique strategy for the synthesis of the highly stable, unusual charge-separated pyridinium-, isoquinolinium-, quinolinium-, and *N*-methylimidazolium-tetronic acid zwitterions from the three-component reaction of pyridine, isoquinoline, quinoline, and *N*-methylimidazole with dialkyl acetylenedicarboxylates and 3-chlorotetronic acid in EtOH at room temperature.

References

Shaabani A, Maleki A, Moghimi-Rad J. (2007) A novel isocyanide-based three-component reaction: synthesis of highly substituted 1,6-dihydropyrazine-2,3-dicarbonitrile derivatives. J Org Chem 72:6309–6317.

Shaabani A, Maleki A, Mofakham H, Moghimi-Rad J. (2008) A novel one-pot pseudo-five-component synthesis of 4,5,6,7-tetrahydro-1*H*-1,4-diazepine-5-carboxamide derivatives. J Org Chem 73:3925–3931.

Shaabani A, Maleki A, Mofakham H. (2008) Novel multicomponent one-pot synthesis of tetrahydro-1*H*-1,5-benzodiazepine-2-carboxamide derivatives. J Comb Chem 10:595–604.

Shaabani A, Soleimani E, Maleki A. (2007) One-pot three-component synthesis of 3-aminoimidazo [1,2-a]pyridines and -pyrazines in the presence of silica sulfuric acid. Monatsh Chem 138:73–76.

Shaabani A, Farhangi E, Rahmati A. (2006) Synthesis of tetrahydrobenzimidazo[1,2-b]quinazolin-1(2H)-one and tetrahydro-1,2,4-triazolo[5,1-b]quinazolin-8(4H)-one ring systems under solvent-free conditions. Comb Chem High Throughput Screen 9:771–776.

Shaabani A, Maleki A. (2007) Ionic liquid promoted one-pot three-component reaction: synthesis of annulated imidazo[1,2-a]azines using trimethylsilylcyanide. Monatsh Chem 138:51–56.

A. Shaabani (✉)
Department of Chemistry, Shahid Beheshti University, 19396-4716 Tehran, Iran
e-mail: a-shaabani@cc.sbu.ac.ir

Synthesis of 4(5)-Imidazolecarboxylic Acid Derivatives

Nikolai A. Beliaev, Vladimir S. Mokrushin, and Igor V. Paramonov

Products of the well-known multi-component imidazole ring formation reaction (the Weidenhagen method) were studied. Earlier it was known that, depending on the reaction conditions, 1*H*-imidazol-4-ylmethanol (3) [1,2,3,4] or 1-(1*H*-imidazol-4-yl)butane-1,2,3,4-tetrol (5) [4] forms. We have found that regardless of reaction conditions, copper salt 1 always contains imidazoles 2–5 in the following amounts: 2, 2–3%; 3, 16–19%; 4, 5–7%; 5, 6–8%. The decomposition of salt 1 in acid media,

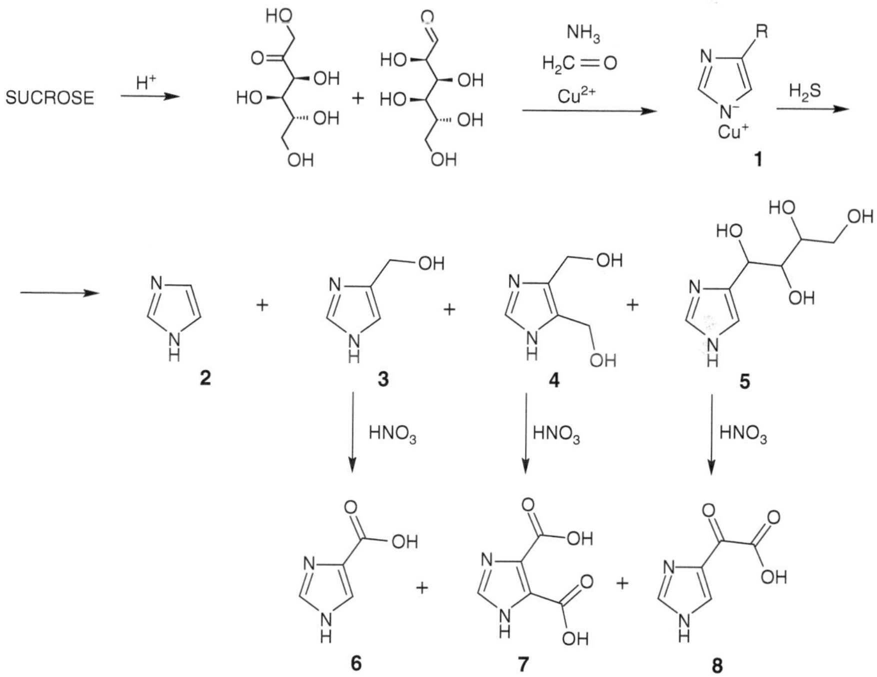

N.A. Beliaev (✉)
Urals State Technical University, Ekaterinburg 620002, Russia
e-mail: beliaev@mail.ustu.ru

the removal of copper by H_2S, and the oxidation of the resulting mixture with diluted nitric acid at a high temperature gave acids 6–8, which were separated based on their different solubility and acidity. These results made it possible to elaborate the preparative synthesis method. After modification, the method was successfully used for the production of acids 6–8 (important intermediates) in kilogram amounts.

References

Weidenhagen R, Herrmann R, Wegner H. (1937) Über neue Abkömmlinge des Imidazols (IV. Mitteil. über Imidazole). Chem Ber 70:575.

Totter JR, Darby WJ. (1944) 4(5)-hydroxymethylimidazole hydrochloride. Organic Synth 24:64.

Darby WJ, Lewis HB, Totter JR. (1942) The preparation of 4(5)-hydroxymethylimidazole. J Am Chem Soc 64:463.

Parrod J. (1932) Bull Soc Chim Fr 51(4):1424–1426.

Parrod J. (1933) Ann Chim (Cachan, Fr) 19(10):205, 232, 233, 238, 245.

Short Communications: Synthesis Reactions

Polyfunctional U-MCR Reagents: Convenient Pathway to Novel Heterocyclic Systems

Alexey Ilyin

U-4MCR is a well-known synthetic method for the synthesis of different polyfunctional organic molecules. The use of bi- or tri-functional reagents usually leads to ring systems, some of which cannot be synthesized by other methods.

Some unsaturated carbonyl compounds in combination with oxy-, amino-, or mercaptoacids are suitable in U-4MCR with spontaneous postcondensation cyclization:

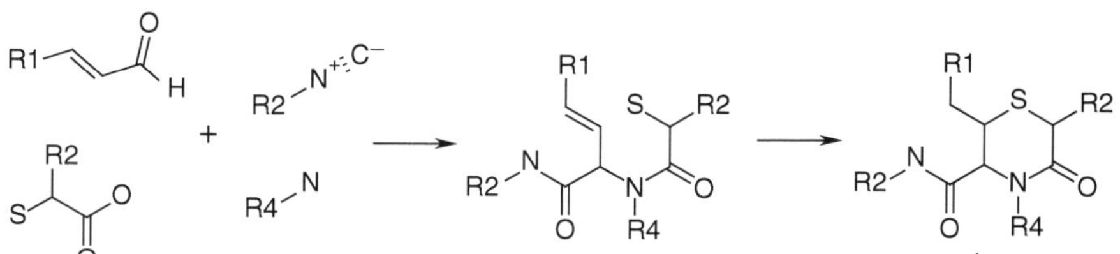

1

Also, the usage of aldehydo- or ketoacids in U-3MCR is convenient for the synthesis of different heterocyclic systems:

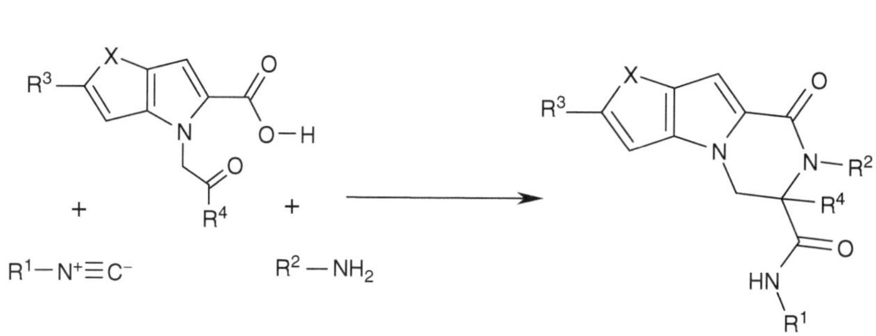

A. Ilyin (✉)
C hemical Diversity Research Institute, Khimki, Moscow Reg, 114401, Russia
e-mail: api@chemdiv.com

Also, interesting postcondensation skeleton rearrangements of tandem U-4MCR products lead to natural-like tricyclic products:

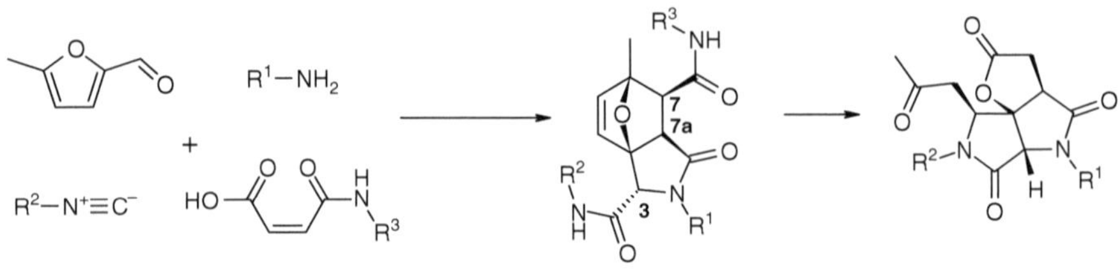

One-Pot Synthesis of Pyridines by Cyclocondensation of Acetyl- and Benzoylpyruvates, Trimethyl Orthoformate, and Enamines

Anna K. Garkushenko, Maria A. Dushek, Galina P. Sagitullina, and Reva S. Sagitullin

The one-pot synthesis of nonsymmetrical pyridines by cyclocondensation of acetyl- (benzoyl)pyruvic acid esters, trimethyl orthoformate, and various enamines has been developed (Scheme 1).

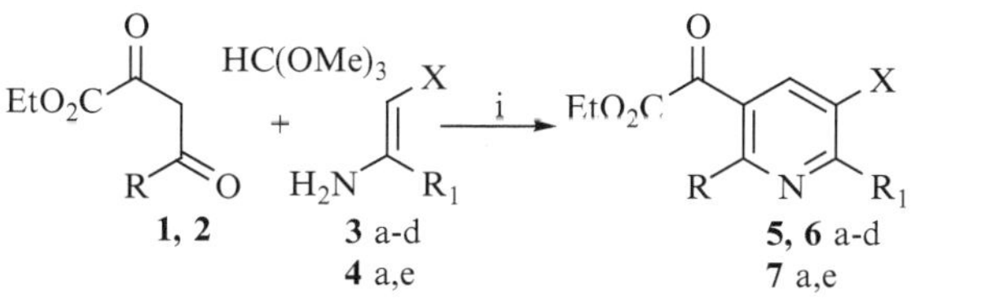

1 R = Me, **2** R = Ph; **3** R_1 = Me, **4** R_1 = Ph;
5 R = R_1 = Me, **6** R = Ph, R_1 = Me, **7** R = Me, R_1 = Ph

a X = CN, **b** X = COMe, **c** X = COPh, **d** X = 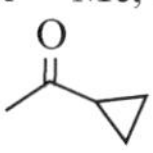, **e** X = NO_2

Scheme 1 One-pot synthesis of pyridines

The synthesis scheme to prepare pyridines **5–7** comprised of diketoesters' **1**, **2** reaction with trimethyl orthoformate results in acetyl-(benzoyl)pyruvate methoxymethylene derivatives. The subsequent conjugated addition of enamines **3**, **4** to the activated double bond of structure **8** (Michael reaction) followed by the two-step elimination of ethanol leads to intermediate **9**. The latter undergoes an intramolecular heterocyclization by the interaction of amino- and acetyl groups to form substituted pyridines **5–7** (Scheme 2).

A.K. Garkushenko (✉)
Department of Organic Chemistry, F.M. Dostoevsky Omsk State University, Mira Prospect 55a, 644077, Omsk, Russia
e-mail: sagitullina@orgchem.univer.omsk.su

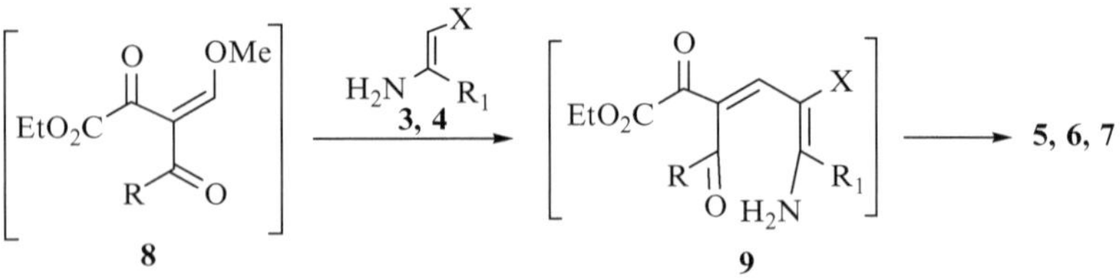

Scheme 2 Mechanism of the reaction

Acknowledgment Financial support of this work by the Russian Foundation for Basic Research (grant no. 07-03-00783-a) is gratefully acknowledged.

Unprecedented Multi-component Approach to 1,4-Diazepine- and 1,4-Benzodiazepine Ring Formation

Volodymyr Kysil, Alexander Khvat, Sergey Tsirulnikov, Sergey Tkachenko, and Alexandre Ivachtchenko

The isocyanide-based multi-component reaction of primary 1,3-diamines 1 with carbonyl compounds 2 has been developed as a novel approach to the 1,4-diazepine ring formation. The reaction of 1,3-diaminopropane **1a** and 2-aminobenzylamine **1b** affords hitherto unknown 2-aminoderivatives of 1,4-diazepine **4a** and 1,4-benzodiazepine **4b**, respectively, including those based on their *spiro*-heterocyclic compounds.

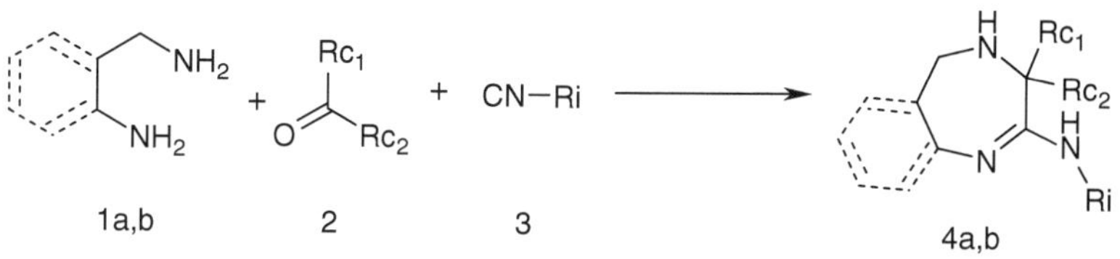

Importantly, the reaction of nonsymmetrical diamine **1b** proved to be highly regioselective, providing only one isomer, **4b**.

Detailed experimental features, mechanisms, and further applications of the developed reaction as well as its scope regarding each component will be discussed.

V. Kysil (✉)
ChemDiv Inc, 11558 Sorrento Valley Rd., #5, San Diego, CA 92121, USA
e-mail: vkysil@chemdiv.com

Synthesis and Interaction with Butylamine of Methyl 6-Acyl-7-Phenyl-4,7-Dihydrotetrazolo[1,5-*a*]Pyrimidine-5-Carboxylates

Vladimir L. Gein and Olga S. Panova

Multi-component reactions have attracted an enormous interest due to their efficacy for the creation of heterocyclic compounds in a single synthetic step. Among the multi-component reactions, the three-component processes have been developed into useful organic procedures. The 1-*H*-1,2,4-tetrazole derivatives are important due to their therapeutic and pharmacological properties.

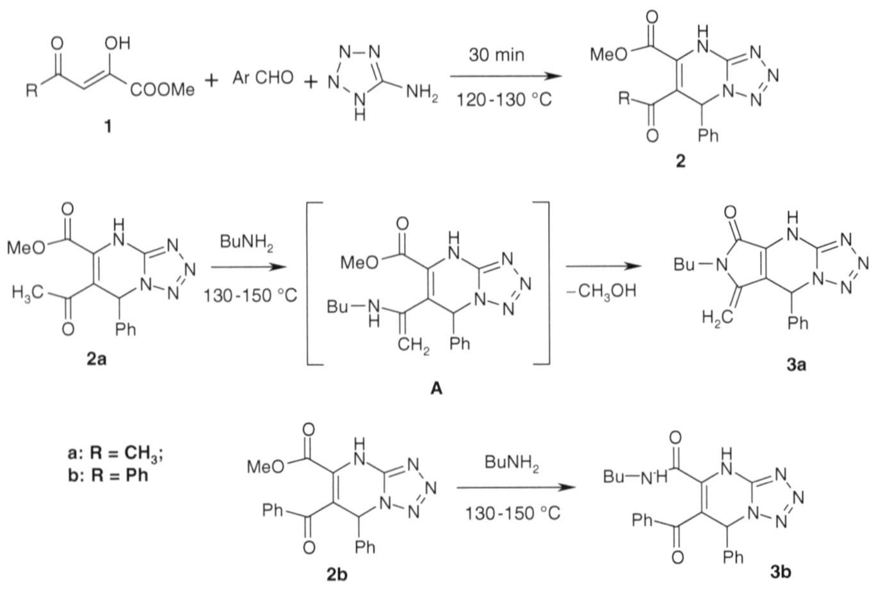

O.S. Panova (✉)
Department of Physical and Colloidal Chemistry, Perm State Pharmaceutical Academy, 48 Lenin Street, 614990, Perm, Russia
e-mail: perm@pfa.ru

Reaction of methyl esters acylpyruvic acids, 5-aminotetrazole, and benzaldehyde leads to the formation of corresponding methyl 6-acyl-7-phenyl-4,7-dihydrotetrazolo[1,5-*a*]pyrimidine-5-carboxylates (**2a**, **b**) under solvent-free conditions. As the next step, to investigate chemical properties, we studied the interaction of heterocyclic compounds **2a**, **b** with butylamine.

It has been observed that the product derived from the reaction of methyl 6-acyl-7-phenyl-4,7-dihydrotetrazolo[1,5-*a*]pyrimidine-5-carboxylates (**2a**, **b**) with butylamine depends on the substitute in the sixth position of the heterocyclic compound. In the case of the acetyl group, in the sixth position, the 6-butyl-7-methylen-8-phenyl-6,7-dihydro-4 -pyrrolo[3,4-*d*]tetrazolo[1,5-*a*]pyrimidin-5(8)-one (**3a**), whose structure was confirmed by X-ray crystallography, is formed. If the sixth position is occupied by a benzoyl radical, the *N*-butyl-6-benzoyl-7-phenyl-4,7-dihydrotetrazolo[1,5-*a*]pyrimidin-5-carboxamide (**3b**) is obtained.

We propose the following mechanism of formation of 6-butyl-7-methylen-8-phenyl-6,7-dihydro-4 -pyrrolo[3,4-*d*]tetrazolo[1,5-*a*]pyrimidin-5(8)-one (**3a**). At the first stage, the intermediate amine derivative A is formed. Then its enamine form is cyclizated into compound **3a**.

Structures of **2a**, **b** and **3a**, **b** were established by elemental analysis, IR and ^{1}H NMR spectroscopic data.

One-Pot Synthesis of 5-Oxo-1,3,4,5-Tetrahydro-Pyrrolo[4,3,2-*de*]Isoquinoline-3-Carboxamides

Alexey P. Ilyn, Dmitri V. Kravchenko, Victor V. Potapov, and Alexandre V. Ivachtchenko

Fused tricyclic systems containing heteroaromatic moiety with different substituents' profiles are in the top list of privileged heterocyclic scaffolds with pronounced pharmacological worth. However, due to a nontrivial structural composition, such compounds have still not found a robust and convenient synthetic decision for the preparation of the focused combinatorial libraries of high diversity. Therefore, the development of effective synthetic strategies to this type of heterocyclic compounds is one of the key objectives in modern combinatorial chemistry.

In the present work we have focused specifically on broadening the scope and synthetic potential of the modified Ugi MCR that we have recently developed and comprehensively evaluated. This synthetic strategy is based fundamentally on the use of bifunctional reagents in Ugi-type condensation. Thus, the combinatorial library of rare 5-oxo-1,3,4,5-tetrahydro-pyrrolo[4,3,2-*de*]isoquinoline-3-carboxamides was obtained using a modified Ugi four-center, three-component reaction (U-4C-3CR). During the initial stage of our synthetic strategy, we obtained a bifunctional aldehydo-acid **1** based on the reaction of methyl 1*H*-indole-4-carboxylate with $POCl_3$ in DMF followed by alkali hydrolysis of the intermediate ester. We have further found that the reaction of 3-formyl-1*H*-indole-4-carboxylic acid **1** with primary amines **2** and isonitrile **3** in methanol at room temperature or 50–60°C for 12–24 h directly led to novel carboxamide derivatives of 5-oxo-1,3,4,5-tetrahydro-pyrrolo[4,3,2-*de*]isoquinolines **4**. The process presumably follows the same initial course as the classical Ugi condensation with an intermediate Schiff base being attacked by the isonitrile to give a nitrilium intermediate, which then undergoes intramolecular cyclization. As proof of the mechanism suggested, we have not observed the formation of both the classical Passerini and branched Ugi-type products. Based on sufficient analytical data (^{1}H NMR, LC–MS, as well as HR–MS analysis), we have unhesitatingly concluded that tricyclic product **4** was properly formed under the applied conditions. Therefore, the developed MCR may

A.V. Ivachtchenko (✉)
ChemDiv, Inc, 6605 Nancy Ridge Dr, San Diego, CA, 92121, USA
e-mail: av@chemdiv.com

provide a valuable practical tool for the synthesis of novel physiologically active agents containing the title core fragment.

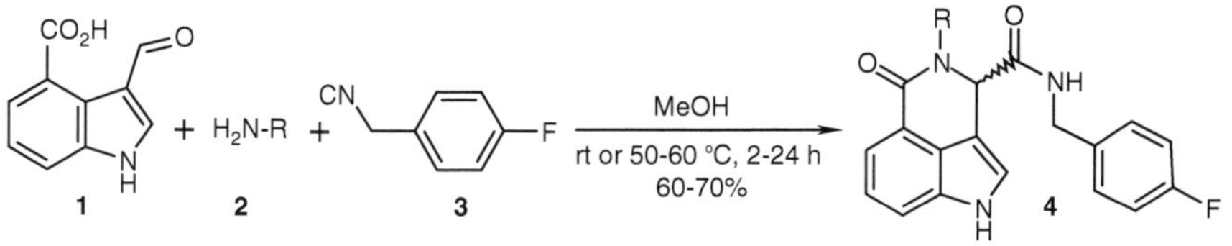

Synthesis of 4-Acyl-5-Aryl-1-(2,2-dimethoxyethyl)-3-Hydroxy-3-Pyrrolin-2-Ones

V. N. Vichegjanina, E. B. Levandovskaya, V. L. Gein, and M. I. Vahrin

It is known that derivatives of tetrahydropyrrol-2,3-diones have expressed antimicrobial activity. With the purpose of synthesizing new tetrahydropyrrol-2,3-diones and studying the influence of their structure on antimicrobial action, it was of interest to enter the 2,2-dimethoxyethyl group into position 1 of the heterocycle.

With this purpose in mind, we investigated the interaction of methyl ether-substituted pyruvic acids with a mix of aromatic aldehyde and 2,2-dimethoxyethanamine. As previous research has shown, the reaction proceeds in dioxane at room temperature with the formation of 4-acyl-5-aryl-1-(2,2-dimethoxyethyl)-3-hydroxy-3-pyrrolin-2-ones (**I–III**).

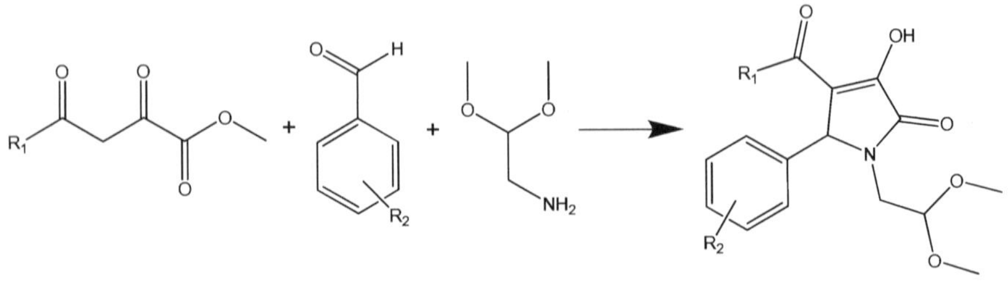

Ia-c, IIa-c, IIIa-h

R^1= CH3 (I); R^2=H (**Ia**), 4-NO_2 (**Ib**); 3-OH (**Ic**);

R^1= C6H5 (II); R^2=H (**IIa**), 4-NO_2 (**IIb**); 3-OH (**IIc**);

R^1= C6H5 (III); R^2=H (**IIIa**), 4-NO_2 (**IIIb**); 3-OH (**IIIc**), 2-Cl (**IIId**), 4-OC_2H_5 (**IIIe**), 4-OH (**IIIf**), 3-NO_2 (**IIIg**), 4-Cl (**IIIh**)

V.N. Vichegjanina (✉)
Perm State Pharmaceutical Academy, Lenin Street, 48, Perm, 614990, Russia
e-mail: perm@pfa.ru

The received compounds (**I–III**) are colorless crystal substances that are soluble in ethanol, acetone, and dimethyl sulfoxide and insoluble in water.

The structure of all the received compounds is established on the basis of data ^{1}H NMR spectroscopy. In ^{1}H NMR spectra of compounds, the group of signals of aromatic protons as a multiplet at 7.2–7.8, a singlet of a proton C5H at 5.18–5.45, signals of protons 1-dimethoxyethyl substituent: a singlet of six protons of two methoxy groups at 3.11–3.22, a multiplet of a proton C1H at 4.35–4.44, a multiplet of a proton C1HaHb at 2.47–2.53, a multiplet of a proton C1HaHb at 3.65–3.72 are present.

All the received compounds (**I–III**) form with an alcoholic solution of FeC13; their intense cherry color testifies to their existence in enolic form.

Synthesis and Structure of Isopropyl 5-Methyl-7-aryl-4,7-Dihydrotetrazolo[1,5-a] Pyrymidine-6-Carboxylates

I. N. Vladimirov, A. A. Zorina, N. V. Nosova, V. L. Gein, O. V. Fedorova, and M. I. Vahrin

It was previously established that the interaction of ethers of ketoacids with a mix of aromatic aldehyde and 5-aminotetrazole or 3-amino-1,2,4-triazole leads to condensed heterocyclic systems. With the purpose of receiving new alkyl 1,5-dihydrotetrazolo[1,5-*a*]pyrimidine-6-carboxylates, we have been investigating the interaction of isopropylic ethers of acetoacetic acids with a mix of 5-aminotetrazole and aromatic aldehyde.

Our research has shown that heating equimolar quantities of initial reagents at 130–170°C for 20–30 min leads to formed 5-methyl-7-aryl-4,7-dihydrotetrazolo[1,5-*a*] pyrimidine-6-carboxylic acid isopropyl esters (**I–VII**).

The received substances (**I–VII**) are colorless or painted crystal substances, soluble in dimethylsulfoxyde and dimethylformamide, and insoluble in ethanol and water.

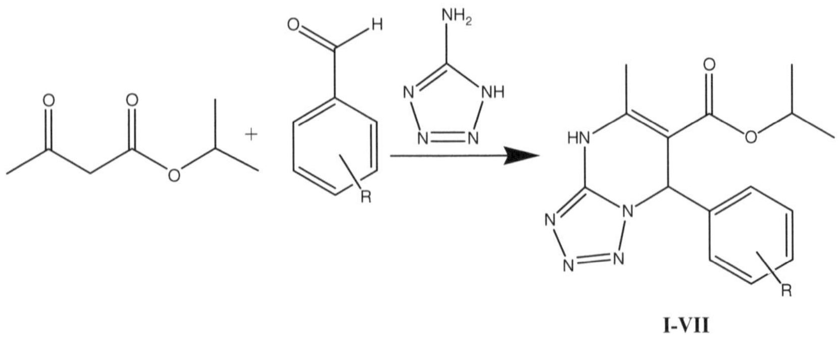

I-VII

R=H(**I**); 3-F (**II**); 4-F (**III**); 4-OH (**IV**); 4-CH_3O (**V**); 3,4-$(CH_3O)_2$ (**VI**); 4-OH-3-OC_2H_5 (**VII**)

I.N. Vladimirov (✉)
Perm State Pharmaceutical Academy, 48 Lenin Street, Perm, 614990, Russia
e-mail: perm@pfa.ru

The structure of all the received compounds (**I–VII**) is established on the basis of data ^1H NMR and infrared spectroscopy. In a nuclear magnetic resonance ^0H spectra of compounds (**I–VII**), except for signals of alkoxygroups, there is the group of signals of aromatic protons as a multiplet at 6.50–6.90, a signal of the proton C5H as a singlet in the field of 6.50–6.70, a signal of the NH-proton at 11.3–11.4, and signals of the CH_3 group as a singlet in the field of 2.25–2.45. In IR spectra the shifts caused by valent fluctuations of the NH-group in the field of 3,200–3,450 sm^{-1}, ether groups in the field of 1,650–1,750 sm^{-1}, and also a shift of valent fluctuations of compounds C=C in the field of 1,660–1,700 sm^{-1}.

The establishment of the structure of the crystal of 5-methyl-7-phenyl-4,7-dihydro-tetrazolo[1,5-*a*] pyrimidine-6-carboxylic acid isopropyl ester (**I**) has been received by slow crystallization from ethanol. We have carried out its rentgeno-structure research. The received results testify to the full conformity of the offered structure of the compound.

Session 3
Multi-Component Reactions in Drug Discovery

MCRs: Bench to Bedside with "Iterative Efficiency"

Christopher Hulme

This presentation will discuss approaches to enhance the rate of molecular probe discovery and close the growing knowledge gap between chemical and biological space created by recent advances in systems biology. As such, multi-component reactions are advocated as tools to build proprietary compound collections, founded on the central tenets of efficiency in medicinal chemistry: (1) the potential for increased "iterative speed" around the "hypothesis–synthesis–screening" loop and (2) reduced numbers of required iterations for expedited value chain progression. Front-loading collections, in this respect, have afforded several successful "bench-to-bedside" studies with no required intermediate "scaffold hopping." Such examples may be viewed as the original "holy grail" of combinatorial chemistry, now enabled by the exponentially increasing MCR-derived "chemical diversity space" made accessible in recent years.

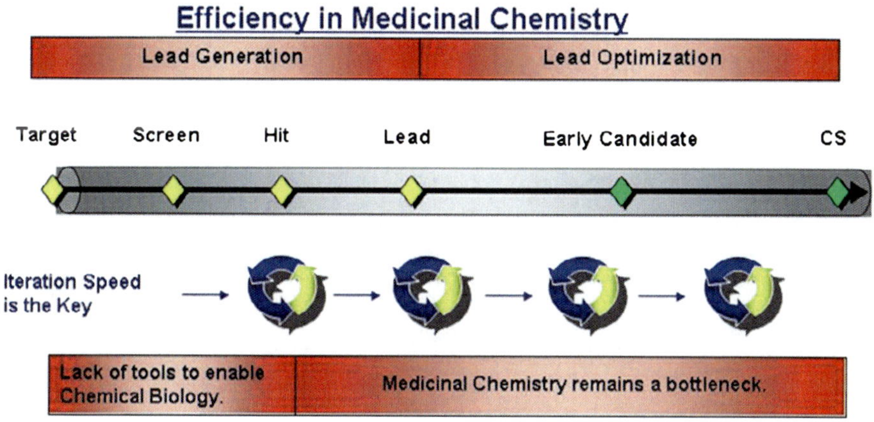

C. Hulme (✉)
University of Arizona, Pharmacy-Skaggs Center, 1703 E. Mabel Street
210207, Tucson, AZ, 85721, USA
e-mail: hulme@pharmacy.arizona.edu

M.A. Mironov (ed.), *MCR 2009*, Advances in Experimental Medicine and Biology 699, DOI 10.1007/978-1-4419-7270-5_3, © Springer Science+Business Media, LLC 2011

Reference

Hulme C, Lee Y-S. (2008) Emerging approaches for the syntheses of bicyclic imidazol[1,2-x]-heterocycles. Mol Div 12:1–15.

Multi-component Reactions and Drug Discovery

Irini Akritopoulou-Zanze

Postmodifications on multi-component reactions have led to the development of easily accessible highly complex structures. The final molecules can be typically obtained in one or two steps employing readily available starting materials. Libraries of compounds can also be prepared for biological screening employing rule of 4 and shape filters. In this presentation we will cover some of our recent work in post-MCR reactions and we will also demonstrate the utility of MCRs for drug discovery.

- Access to novel molecules via combinations of MCRs with other reactions
- First-step MCR: Ugi, van Leusen
  - These reactions tolerate a variety of functional groups
- Second-step reaction: Heck, Huisgen triazole synthesis, INOC, RCM, C–H activation, photochemistry, etc.
  - Intramolecular versions provide novel analogs
- Third-step reaction: Heck, etc.
  - Higher complexity achieved.

Reference

Akritopolou-Zanze I. (2008) Isocyanide-based multicomponent reactions in drug discovery. Curr Opin Chem Biol 12:324–331.

I. Akritopoulou-Zanze (✉)
Abbott Laboratories, Medicinal Chemistry Technologies, 100 Abbott Park Road, Abbott Park, IL 60044, USA
e-mail: Irini.Zanze@abbott.com

Building Molecular Shape Diversity via Post-Ugi Annulations

Wei-Min Dai

In connection with our effort to generate molecular shape diversity starting from readily available 2-aminophenols, we have established highly efficient methodologies for the assembly of novel heterocycles possessing flat, cleft, helical, and spirocyclic shapes. These molecular entities are created, in combination with microwave chemistry, by one-pot Ugi–S_N2 and Ugi–S_NAr sequences followed by

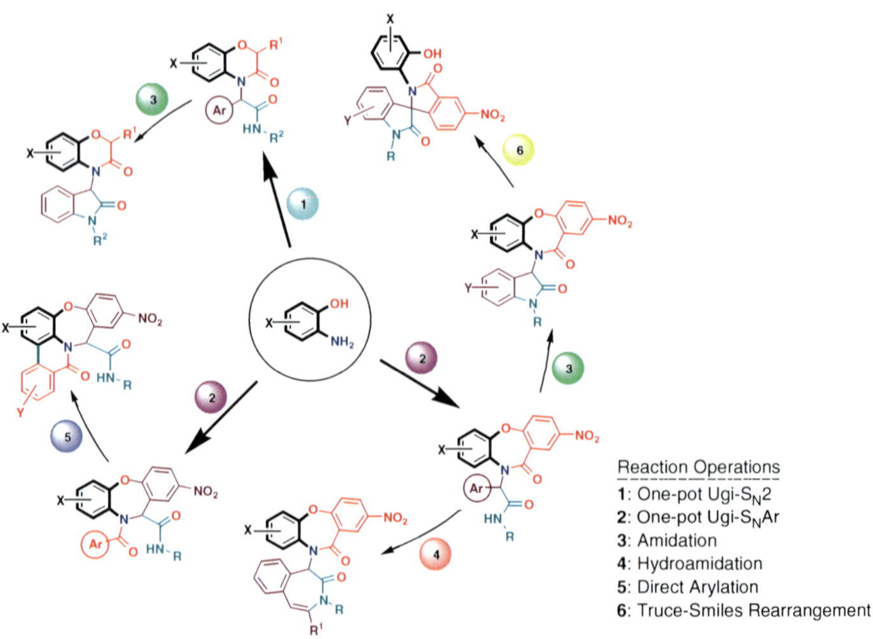

W.-M. Dai (✉)
Laboratory of Asymmetric Catalysis and Synthesis, Department of Chemistry, Zhejiang University, Hangzhou, 310027, China
and
Department of Chemistry, Hong Kong University of Science and Technology, Clear Water Bay, Kowloon, Hong Kong SAR, China
e-mail: chdai@zju.edu.cn

other annulation reactions. The latter include the Pd-catalyzed amidation, hydroamidation, and direct arylation, and a Truce–Smiles rearrangement. The scope of these post-Ugi annulations is presented with a conceptual discussion.

Acknowledgment Financial support from The National Natural Science Foundation of China (grant no. 20672092), Zhejiang University, and Zhejiang University Education Foundation is acknowledged.

References

Dai W-M, Shi J. (2007) Diversity-oriented synthesis and solid-phase organic synthesis under controlled microwave heating. Comb Chem High Throughput Screen 10:837–856.
Dai W-M et al. (2006) C-N bond-linked conjugates of dibenz[*b*,*f*][1,4]oxazepines with 2-oxindole. Synlett 13:2099–2103.

A Fast and Efficient Approach to Small Molecular Weight Protein-Protein Interaction Antagonists Using MCR Chemistry

Barbara Beck, Stuti Srivastava, Wei Wang, Kan Wang, Anna Czarna, Tad A. Holak, Lidio Meireles, Carlos Camacho, Balu Raghavan, Billy W. Day, and Alexander Dömling

Protein–protein interactions (PPIs) have recently become major important drug discovery targets. However, due to the variability of PPI interfaces, no rational and general approach to the discovery of small molecular weight antagonists has been described so far. The interfaces of PPIs are characterized by areas of water exclusion and tight arrays of hydrophobic, hydrogen bonding and charge–charge interactions. Often PPIs comprise so-called hot spots, which are primarily responsible for the energy of this interaction. Thus, small molecular weight compounds have to compete with a dense array of interactions in order to antagonize PPIs efficiently.

We hypothesize that the chemical space of small molecular weight compounds based on multi-component reaction (MCR) chemistries is better suited to discover PPI antagonists as compared to other chemical space. This is based on the much denser assembly of atoms and functional groups in MCR molecules as opposed to non-MCR molecules. However, an issue with the rational drug discovery of PPI antagonists using MCR chemistry is the extremely large and diverse chemical space, which is not suitable for exhaustive experimental approaches.

Therefore, we present here a first and general rational drug discovery approach to PPI antagonist based on a tight interplay of the "anchor concept," virtual and real MCR chemistry, docking, and high content screening (Fig. 1).

A. Domling (✉)
Departments of Pharmacy and Chemistry, Drug Discovery Institute, University of Pittsburgh
3501 5th Avenue, Pittsburgh, PA 15261, USA
e-mail: asd30@pitt.edu

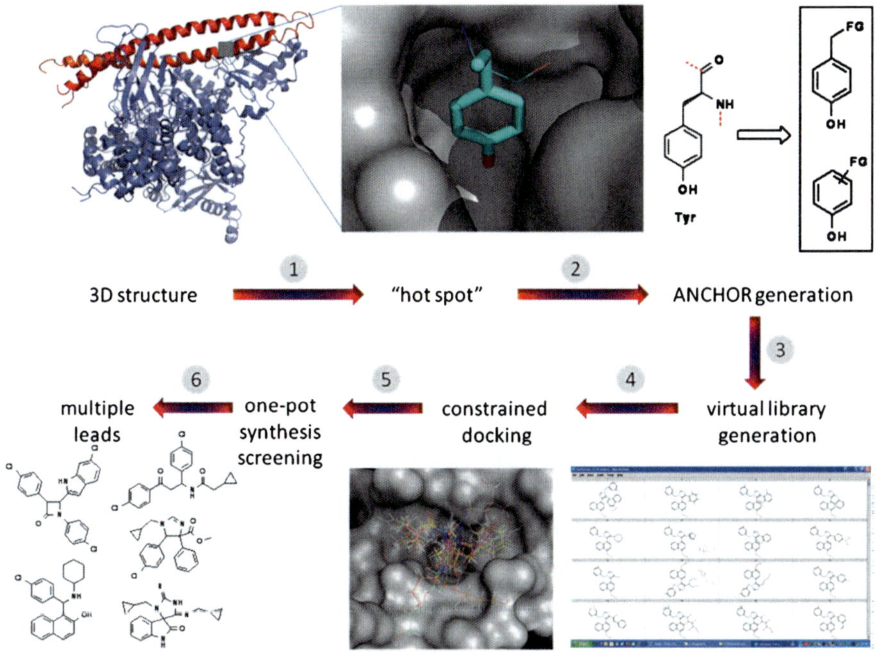

Fig. 1 Docking analysis for PPI antagonists

References

Dömling A. (2006) Recent developments in isocyanide based multicomponent reactions in applied chemistry. Chem Rev 106:17–89.

Czarna A, Beck B, Srivastava S, Popowicz GM, Wolf S, Huang Y, Bista M, Holak TA, Dömling A. (2010) Robust generation of lead compounds for protein-protein interactions by computational and MCR chemistry: p53/Hdm2 antagonists. Angew Chem Int Ed Engl 49(31):5352–5356.

Popowitz G et al. (2010) Cell Cycle 6:1104.

Camacho C, Dömling A, et al. Submitted.

Molecular Diversification and Solid-Phase Synthesis of *N*-Acylpeptides for Antibiotic Screening

Chuanguang Qin, Ruijie Zhang, Qiuyu Wang, Jin Ren, and Linqi Tian

The emergence in all regions of the world of strains of pathogenic bacteria and fungi with resistance to commonly used antibiotics constitutes a potentially serious threat to public health and has necessitated a search for novel types of antimicrobial agents to which the microorganisms have not been exposed. During the last decade, relatively few new antibiotics have been introduced by the pharmaceutical industry, and none has improved activity against multidrug-resistant Gram-negative bacteria. Acylpeptides are unique among the antimicrobial peptides in that they are relatively smaller in size as compared to other cationic antimicrobial peptides and yet exhibit excellent antimicrobial activity. Fatty acid acylation of antimicrobial peptides of bacterial and fungal origin have mostly been limited to non-gene-encoded peptides from *Pseudomonas syringae*. Consequently, several studies to assess the effect of acylation have revealed that conjugation of fatty acids to cationic peptides enhances their antimicrobial activity. However, it is not clear if acylation influences the association of hydrophilic portions of peptides with membranes or modulates the orientation of peptides in membranes. Therefore, it would be relevant to study the interactions of antimicrobial acylpeptides and anionic model membranes as it would help in understanding the biophysical properties governing the association of fatty acylated peptides with biological membranes. Here, we designed and synthesized six fatty *N*-acylated peptides (see the synthesis scheme below) for screening antibiotics against clinical drug-resistant microbes. The results indicated that N-palmitylpeptide, C_{16}-KGGKGGKGGK-NH_2 (4) among the six synthetic acylpeptides, possesses the best antimicrobial activities against clinical multidrug-resistant bacteria such as methicillin-resistant *Staphylococcus aureus* (Gram-positive) and quinolone-resistant *Escherichia coli* (Gram-negative).

C. Qin(✉)
Faculty of Life Science, Northwestern Polytechnical University,
Xi'an, 710072, China
e-mail: qinchg@nwpu.edu.cn

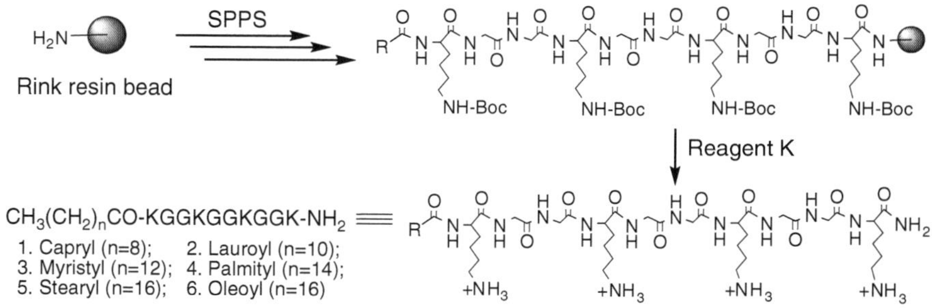

References

Qin CG et al. (2003) Dissociation of antibacterial and hemolytic activities of an amphipathic peptide antibiotic. J Med Chem 46:4830–4833.

Ghielmetti M et al. (2005) Synthetic bacterial lipopeptide analogs: structural requirements for adjuvanticity. Immunobiology 210:211–215.

Barchiesi F et al. (2007) *In vitro* activity of the synthetic lipopeptide PAL-Lys-Lys-NH(2) alone and in combination with antifungal agents against clinical isolates of *Cryptococcus neoformans*. Peptides 28:1509–1513.

Convertible Products of Ugi Reaction Involving Azole Carboxylic Acids and *tert*-Butyl Isocyanide

Mikhail Nikulnikov and Mikhail Krasavin

Having discovered an unusual convertible behavior of Ugi reaction products **1** formed of 1*H*-pyrazole-3-carboxylic acid and *tert*-butyl isocyanide components, we set out to establish whether these findings could be extended to other isocyanides. The results of these studies will be discussed (Scheme 1).

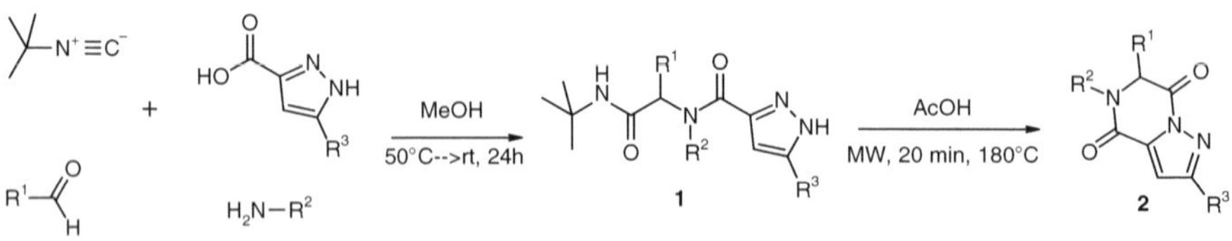

Scheme 1 Synthesis of pyrazoles via MCR

Additionally, we discovered that the (formally) internal pyrrazolides **2** are reactive toward a range of nucleophiles, thus exhibiting the greater utility of this new and technically simple transformation (Scheme 2).

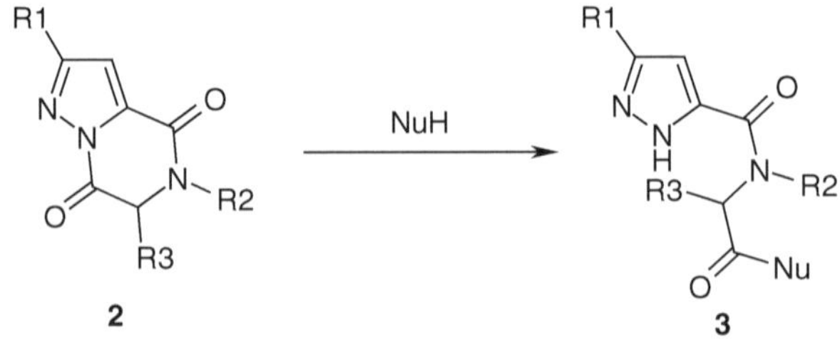

Scheme 2 Derivatization of pyrazole scaffold

M. Krasavin (✉)
Chemical Diversity Research Institute, 2a Rabochaya Street,
Khimki, Moscow Reg, 141400, Russia
e-mail: myk@chemdiv.com

Finally, we are in the process of expanding these findings onto other azole carboxylic acids. Preliminary general indications of this process reaching beyond 1*H*-pyrazole-3-carboxylic acids will be presented.

Reference

Nikulnikov M, Tsirulnikov S, Kysil V, Ivachtchenko A, Krasavin M. (2009) *tert*-Butyl isocyanide as a convertible reagent in Ugi reaction: microwave-assisted preparation of 5,6-dihydropyrazolo[1,5-*a*] pyrazine-4,7-diones. Synlett 20(2):260–262.

Applications of Quinoxaline Synthesis via Isocyanide-Based MCR

Mikhail Krasavin, Vladislav Parchinsky, Sergey Shkavrov, and Konstantin Bukhryakov

Last year we described a new synthesis of quinoxalines via a multi-component reaction among aromatic 1,2-diamines, aldehydes, and isocyanides and subsequent DDQ oxidation of the intermediate redox-unstable 1,4-dihydroquinoxalines (Scheme 1).

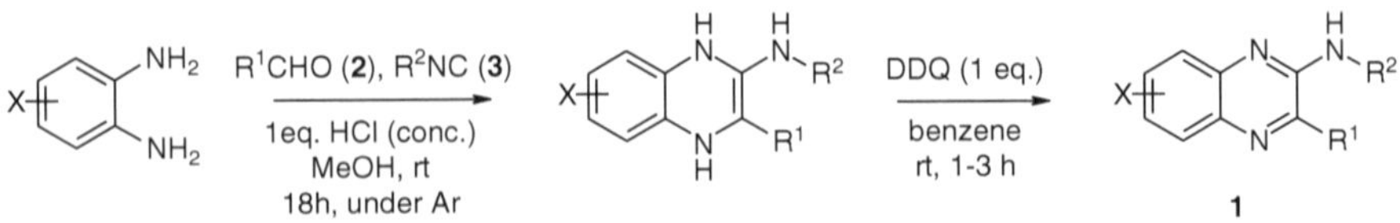

Scheme 1 Synthesis of quinoxalines via MCR

While for the majority of simple products **1**, alternative syntheses have been described, the experimental simplicity and higher functional group tolerance of this two-step, one-purification protocol prompts us to investigate possible applications of this new route to quinoxalines. Our recent results in this direction will be summarized. In particular, the recently established utility of the above MCR process in an MCR/MCR coupling strategy will be presented (Scheme 2).

M. Krasavin (✉)
Chemical Diversity Research Institute, 2a Rabochaya Street, Khimki, Moscow Reg, 141400, Russia
e-mail: myk@chemdiv.com

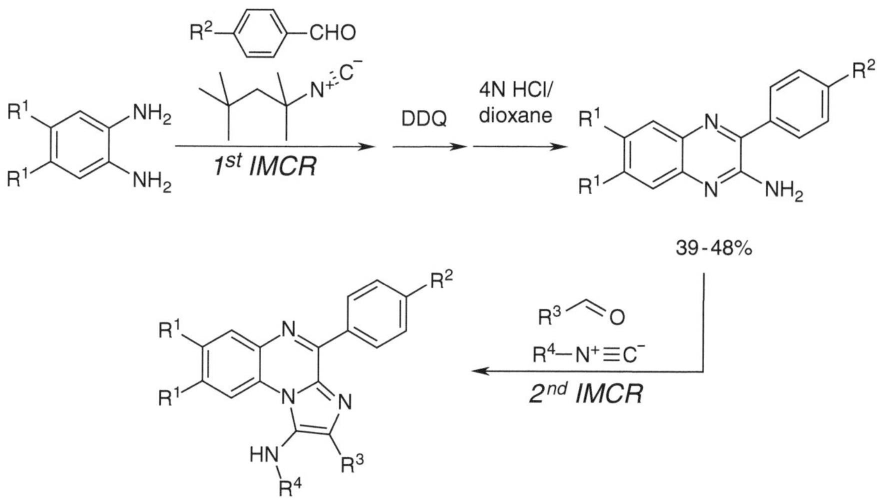

Scheme 2 An example of MCR/MCR coupling strategy

Reference

Krasavin M, Parchinsky V. (2008) Expedient entry into 1,4-dihydroquinoxalines and quinoxalines via a novel variant of isocyanide-based MCR. Synlett 2008:645–648.

Continued Evidence for the Utility of TMSCl as Promoter for Groebke-Blackburn Reactions: Synthesis of Imidazo[2,1-*b*][1,3]benzothiazoles

Sergey Tsirulnikov and Mikhail Krasavin

We have recently established that equimolar TMSCl efficiently promotes Groebke–Blackburn multi-component reactions (GB-MCRs) of 2-aminoazines and -azoles, often for "difficult" substrates that otherwise react poorly in these reactions, thus opening a broader substrate landscape for this atom-economical process. We continue to study the applicability of these conditions to other GB-MCRs, aiming to establish this as a method of choice. Recently we demonstrated that TMSCl also promoted GB-MCR of 2-aminobenzothiazoles, providing good to excellent yields of regiochemically unambiguous imidazo[2,1-*b*]benzothiazoles (Scheme 1).

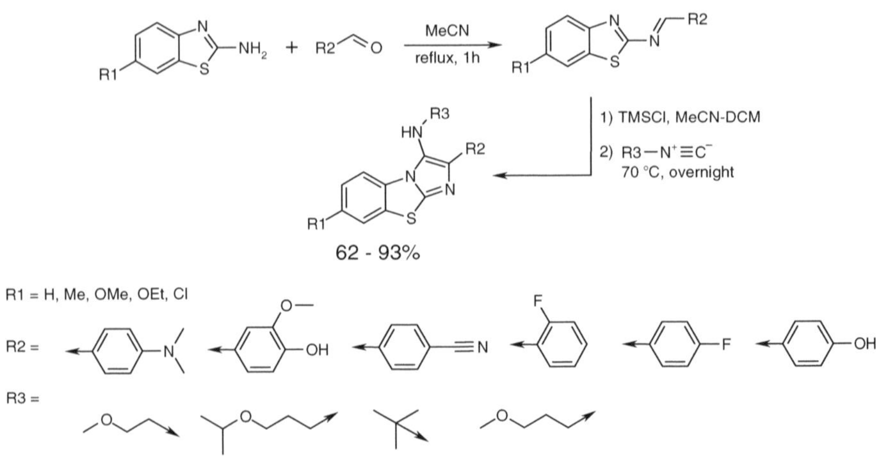

Scheme 1 Use of 2-aminobenzothiazoles in MCR

M. Krasavin (✉)
Chemical Diversity Research Institute, 2a Rabochaya Street, Khimki, Moscow Reg, 141400, Russia
e-mail: myk@chemdiv.com

While this reaction is also workable under NH_4Cl/toluene reflux conditions that we introduced, the present findings clearly extend the utility of TMSCl protocols and speak for the universal character of this promoter as far as GB-MCR is concerned.

References

Krasavin M, Tsirulnikov S, Nikulnikov M, Kysil V, Ivachtchenko A. (2008) *tert*-Butyl isocyanide revisited as a convertible reagent in the Groebke–Blackburn reaction. Tetrahedron Lett 49:5241–5243.

Parchinsky V, Shuvalova O, Ushakova O, Kravchenko DV, Krasavin M. (2006) Multi-component reactions between 2-aminopyrimidine, aldehydes and isonitriles: the use of a nonpolar solvent suppresses formation of multiple products. Tetrahedron Lett 47:947–951.

Systematic Diversity-Oriented Synthesis Based on Ugi Reaction Products

Ekaterina Bushkova, Vladislav Parchinsky, and Mikhail Krasavin

Diversity-oriented synthesis (DOS) in its conceptually "pure" form requires such a design of reaction arrays as would allow applying the same sequence of chemical events to individual combinations of reagents, ultimately arriving at a product set where each of the products would represent a unique scaffold. Thus, with all elements of a combinatorial chemistry mentality (reagent diversity with redundant synthetic steps), such an approach would yield a *library of scaffolds*, where each compound singly addresses considerably different parts of the chemical space.

Post-Ugi modifications have served many researchers well in helping elaborate the linear petoid Ugi reaction product *template* into many intriguing scaffolds. We recognized that products of Ugi reaction of acylhydrazones with isonitriles and TFA (**1**) contain a reactive center that can be used efficiently for numerous post-Ugi modifications. With the proper design of these transformations, various aza-peptoid templates **1** (already including three elements of R group diversity) can yield a different template, thus systematically realizing the concept of DOS. Such transformations that are currently tested in our laboratories may include (1) second isocyanide-based MCR (yielding tripeptoid scaffolds **2** with five (!) elements of diversity); (2) acylation of the reactive nitrogen with maleic anhydride and intramolecular Diels–Alder reaction onto furan moiety (if furan-2-carboxaldehyde was used in the preparation of **1**), resulting in the formation of **3**.

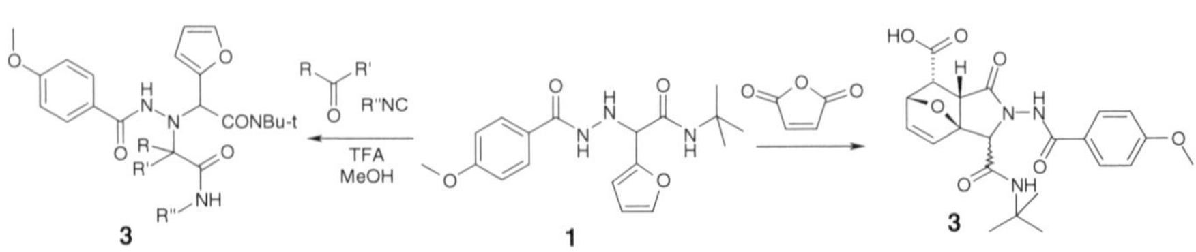

M. Krasavin (✉)
Chemical Diversity Research Institute, 2a Rabochaya Street,
Khimki, Moscow Reg, 141400, Russia
e-mail: myk@chemdiv.com

The results of our latest, off-the-bench findings will be presented and discussed.

Reference

Ilyin A, Kysil V, Krasavin M, Kurashvili I, Ivachtchenko AV. (2006) Complexity-enhancing acid-promoted rearrangement of tricyclic products of tandem Ugi 4CC/intramolecular Diels–Alder reaction. J Org Chem 71:9544–9547.

Short Communications

Priaxon AG, a New Dimension in Drug Discovery

Cédric Kalinski

The goal of this oral presentation is to introduce the MCR-based technology developed by Priaxon AG and to summarize the new MCRs reported by our team in the last 3 years.

Priaxon opens up a new dimension in drug discovery by the combination of several hundreds of validated MCRs with classical organic chemistry, giving access to the largest molecular space of drug-like compounds (>1 billion). Typical applications in medicinal chemistry are drug design or scaffold-hopping projects. By running through a succession of application-specific 1D filters, 2D topological filters and 3D docking-suitable structures are extracted from the terabyte database. The most promising candidates are then refined by conventional molecular modeling. Since the input of structures has already been based on straightforward preparative accessibility and drug-likeness, the selected molecules can then be synthesized with high success rates and subjected to biological testing.

Priaxon can enlarge this molecular space by discovering new MCRs or by investigating new combinations between well-known MCRs and classical postcondensation reactions. In recent years, we have reported the combinatorial synthesis of 4-oxo-4H-imidazo[1,5-*a*]quinoxalines and 4-oxo-4H-pyrazolo[1,5-*a*]quinoxalines (*Tetrahedron Lett*, 2007, 48:8060) and the diversity-oriented synthesis of benzoxazoles and benzothiazoles (*Tetrahedron Lett*, 2007, 48:9030) and illustrated that MCRs could also be a powerful tool for generic drug synthesis (*Synthesis*, 2008, 24:4007).

C. Kalinski (✉)
Priaxon AG, Gmunder Strasse 37-37a, D-81379 München, Germany
e-mail: kalinski@priaxon.com

Isocyanide-Based Multi-component Reaction of Diamines: An Efficient Strategy for Assembling of Unique Heterocyclic Scaffolds

Volodymyr Kysil, Sergey Tsirulnikov, and Alexandre Ivachtchenko

On the one hand, the development of small molecular weight scaffolds that contain privileged heterocyclic cores with a high degree of diversity is a leading focus in medicinal chemistry. On the other hand, isocyanide-based multi-component reactions (IMCRs) have become one of the most valuable tools of drug discovery-oriented synthetic organic chemistry since they allow synthesizing diverse natural-like small molecules in simple one-pot procedures. The presentation will be focused on a recently developed IMCR of various primary diamines and carbonyl compounds that leads to a wide variety of heterocyclic scaffolds with pyrazine, quinaxoline, hetarenopyrazine, 1,4-diazepine, 1,4-benzodiazepine, and other pharmaceutically relevant cores as well as heterocyclic spiranes based on them.

The scope, limitations, and subsequent promising applications of the MCR developed will be discussed.

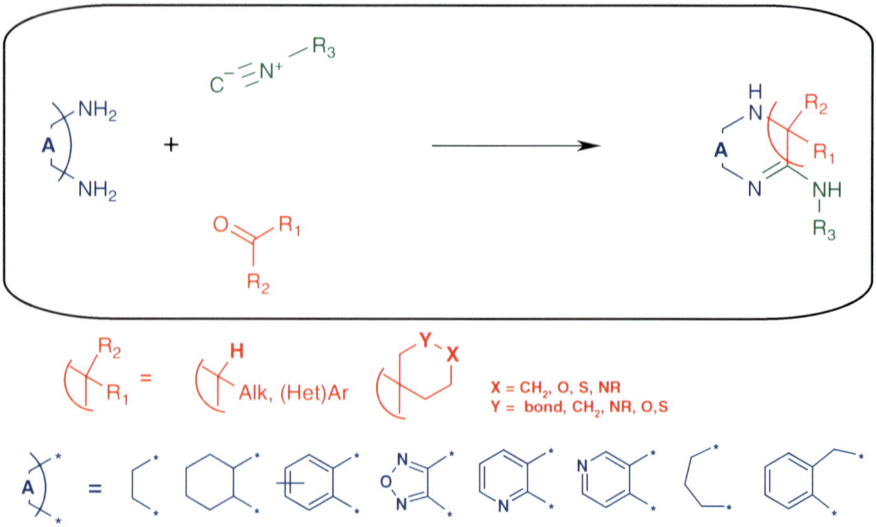

V. Kysil (✉)
ChemDiv Inc, 11558 Sorrento Valley Road, #5, San Diego, CA 92121, USA
e-mail: vkysil@chemdiv.com

A Convenient Synthesis of Novel Heterocyclic Structures Containing 11-Oxo-6,11,12,13-tetrahydrodibenzo[*b*,*g*][1,5]oxazonine Fragment

Victor V. Potapov, Volodymyr M. Kysil, Nataliya A. Fetisova, Alexandr V. Nikitin, and Alexandre V. Ivachtchenko

Among a variety of physiologically active aryl- and heteroaryl-fused six- and seven-membered heterocyclic systems, including morpholines and oxazepines, nine-membered diaryl-fused heterocycles form a relatively little-explored group with a pronounced pharmaceutical importance in terms of the fundamental concept of bioisosterism. In particular, 11-oxo-6,11,12,13-tetrahydrodibenzo[*b*,*g*][1,5] oxazonines and their carboxamide analogs represent promising synthetic targets with a pronounced pharmaceutical importance. Therefore, the development of effective synthetic strategies to this type of heterocyclic compounds is one of the key objectives in modern combinatorial chemistry.

In the present work we have focused specifically on broadening the scope and synthetic potential of the modified Ugi MCR that we recently developed and comprehensively evaluated. This synthetic strategy is based primarily on the use of bifunctional reagents in Ugi-type condensation. Following the developed synthetic approach, we obtained heterocyclic compounds **5** that were not previously described in the thematic literature, using 2-[(2-formylphenoxy)methyl]benzoic acid **4**. The key bifunctional compound **4** was easily obtained from acid **1** and corresponding salicylaldehyde **2** following a modified variant of a known synthetic method. We have further found that the reaction of aromatic aldehyde-acid **4** with primary amines and isonitriles in methanol at 25°C for 5–10 h led to novel 11-oxo-6,11,12,13-tetrahydrodibenzo[*b*,*g*][1,5]oxazonine-13-carboxamides **5**. The assignment of all synthesized structures was made on the basis of ^{1}H NMR and high-resolution mass-spectroscopy data. In some cases pure crystalline substances were obtained, thereby allowing us to reveal the structures of the resulting compounds accurately. Based on the obtained analytical data, it was eventually concluded that the target compounds **5** were obtained as racemic mixtures of corresponding enantiomers.

V.V. Potapov (✉)
Department of Organic Chemistry, Chemical Diversity Research Institute,
114401 Khimki, Moscow Reg, Russia
e-mail: vai@chemdiv.com

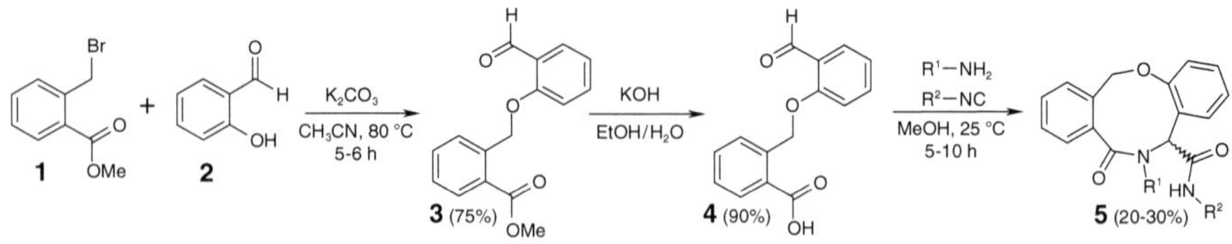
Br
1
OMe
2
OH
H
K2CO3
CH3CN, 80 °C
5-6 h
3 (75%)
KOH
EtOH/H2O
4 (90%)
R1—NH2
R2—NC
MeOH, 25 °C
5-10 h
5 (20-30%)

Synthesis of Substituted 3-Methyl-1-oxo-2,3,4,5-tetrahydro-1H-benzo[C]azepino-3-carboxamides

Victor V. Potapov, Alexey P. Ilyn, Nataliya A. Fetisova, Dmitri V. Kravchenko, and Alexandre V. Ivachtchenko

In this work we present a convenient synthetic approach to novel 3-methyl-1-oxo-2,3,4,5-tetrahydro-1*H*-benzo[*c*]azepines and their carboxamide analogs. The combinatorial library of the title heterocyclic compounds was obtained using a modified Ugi four-center, three-component reaction (U-4C-3CR) that we recently developed and comprehensively validated. The designed approach was primarily based on the use of bifunctional keto-acid as one of the Ugi reactants. The key results of our success within this field were represented in a series of scientific papers dedicated to advanced developments in modern organic chemistry, especially in combinatorial synthesis and MCRs.

In the initial step, the developed synthetic route included the synthesis of bifunctional compounds for a modified Ugi MCR. Thus, the corresponding 2-(3-oxobutyl)benzoic acid **1** was easily obtained from 2-bromomethyl-benzoic acid methyl ester by the alkylation with methyl acetoacetate followed by mild alkali hydrolysis and further decarboxylation. The core heterocyclic fragment **4** was then obtained in methanol medium by the reaction of keto-acid **1** with various primary amines **2** and 3-methyl-butylisonitrile **3**. It was found that the yields of the target cyclic Ugi products depend only on the solubility of amines and isonitrile in the reaction mixture. It should particularly be noted that the purification procedure was successfully adapted for the majority of combinatorial library members. The assignment of all synthesized structures was made on the basis of ^{1}H NMR and HR-MS data. These data gave satisfactory results consistent with the suggested molecular structures. Finally, the synthesized heterocyclic compounds can be reasonably regarded as potential drug compounds with promising pharmacological profiles. Therefore, the developed MCR may provide a valuable, comprehensive, practical tool for the creation of novel physiologically active agents containing 1-oxo-2,3,4,5-tetrahydro-1-*H*-benzo[*c*]azepine fragment.

A.V. Ivachtchenko (✉)
ChemDiv, Inc, 6605 Nancy Ridge Dr, San Diego, CA 92121 USA
e-mail: av@chemdiv.com

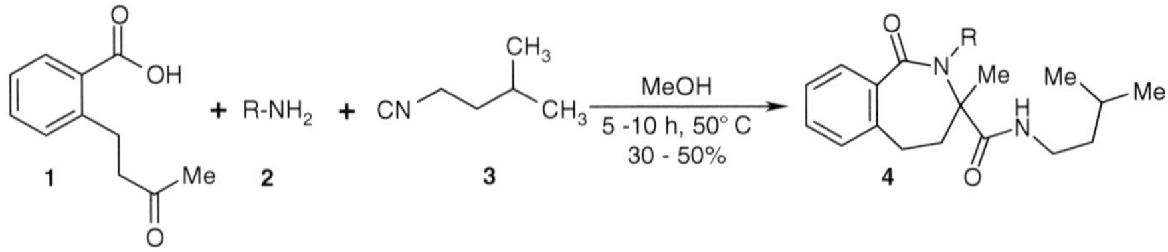
OH
Me
1
+ R-NH2
2
+ CN
CH3
CH3
3
MeOH
5 -10 h, 50° C
30 - 50%
R
N
Me
Me
Me
H
N
4

New Approach of Some Bifunctional Initials in U-4MCR

Olga Shilova, Alexey Ilyin, and Alexandre Ivachtchenko

Tandem reactions in preparative chemistry are useful methods for the synthesis of different heterocycles because of usually high yields and simple experimental procedures. When two or more reagents in Ugi four component (U-4MCR)

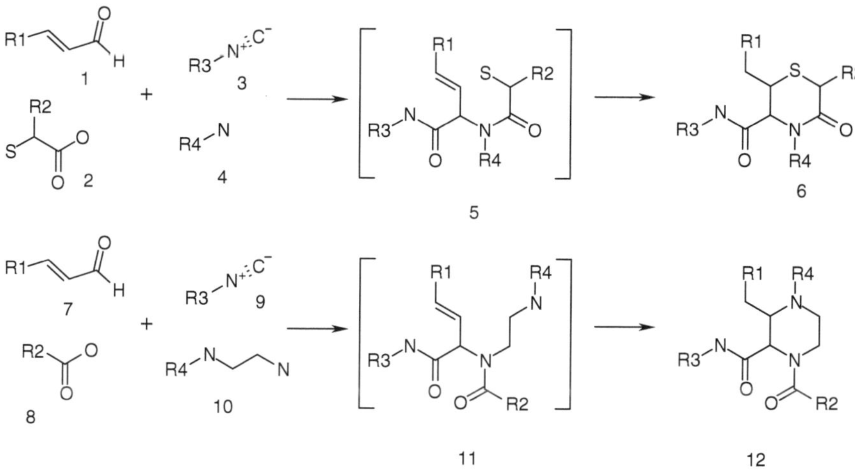

are bifunctional, the initial Ugi product can rearrange or cyclize in the same reaction conditions.

For example, unsaturated aldehydes such as carbonyl reagent in U-4MCR are suitable for some postcondensation transformations. When the second bifunctional initial is α-mercaptoacid, thiomorpholinon derivatives (**6**) can easily be synthesized. Also, diamines (**10**) yield piperazine derivatives (**12**).

A. Ilyin (✉)
Chemical Diversity Research Institute, Khimki, Moscow Reg, 114401, Russia
e-mail: api@chemdiv.com

One-Step Assembly of Novel Carbamoyl-Substituted Heterocycle-Fused 1-Oxo-2,3-dihydro-1*H*-benzo[c]azepines

Alexandre M. Shkirando, Volodymyr M. Kysil, Victor V. Potapov, and Alexandre V. Ivachtchenko

Among the existing pharmacological agents, aryl-fused seven-membered heterocyclic compounds, in particular benzazepines, benzodiazepines, and their bioisosteric analogs, were found to have a wide spectrum of physiological activity. For example, benzazepinone fragments can be found in many therapeutic agents already released onto the pharmaceutical market or currently being evaluated in advanced clinical trials. However, heterocycle-fused benzazepinones are still relatively poorly described in the scientific literature; moreover, comprehensive synthetic approaches toward this type of heterocyclic scaffold with different substituents' profile have not been thoroughly studied.

In this work we present a convenient synthetic route to novel heterocycle-fused 1-oxo-2,3-dihydro-1*H*-benzo[*c*]azepines **3a–d** which were obtained by a modified Ugi four-center, three-component reaction (U-4C-3CR) based on the use of heterocyclic aldehyde-acid as the bifunctional component. Initially, the bifunctional aldehyde-acids **1a–c** were synthesized by a palladium-catalyzed Suzuki reaction, a powerful synthetic strategy that was not applied previously for the preparation of such compounds. The obtained compounds were further used in Ugi MCR with primary amines and isonitrile. We have found that yields and structures of the resulting Ugi products are directly dependent on the structure of the initial components. Particularly, the reaction among compound **1a**, **b**, amine **2b**, **c**, and cyclopentyl isonitrile led only to the stable intermediate Schiff bases, in contrast to the analogous MCR among compound **6a**, **c**, amine **2a**, and the corresponding isonitrile. The intermediate Schiff bases can be successfully converted into the title heterocycle-fused 1-oxo-2,3-dihydro-1*H*-benzo[*c*]azepines by heating the obtained solution to 100°C and keeping it for 1–2 h in a microwave-reactor.

A.V. Ivachtchenko (✉)
ChemDiv, Inc, 6605 Nancy Ridge Dr, San Diego, CA 92121 USA
e-mail: av@chemdiv.com

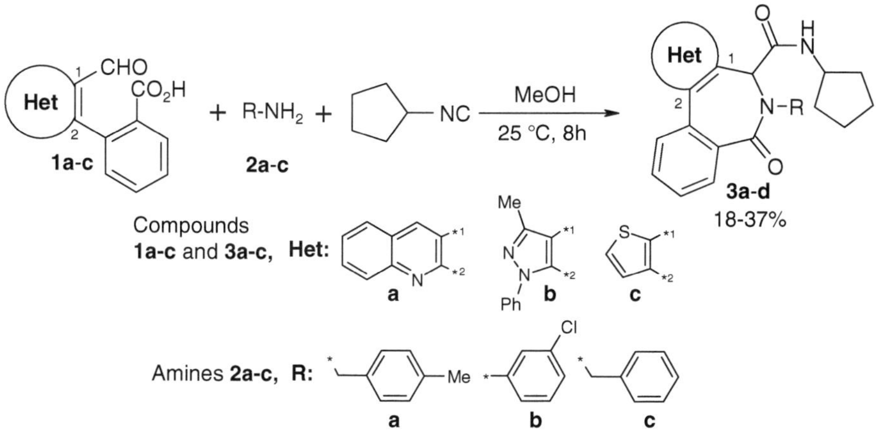
CHO
CO_2H
Het
1a-c
R-NH2
2a-c
NC
MeOH
25 °C, 8h
Het
N-R
3a-d
18-37%
Compounds
1a-c and 3a-c, Het:
a
b
c
Me
Ph
Amines 2a-c, R:
Me
Cl
a
b
c

A Facile One-Pot Greener Synthesis and Antibacterial Activity of 2-Amino-4H-pyran Derivatives

Buchi Reddy Vaddula, Dalip Kumar, Shashwat Sharad, Urvashi Dube, and Suman Kapur

The development of environmentally benign, efficient, and economical methods for the synthesis of biologically interesting compounds remains a significant challenge in synthetic chemistry. One-pot green synthesis emphasizes the need for environmentally clean synthesis, which involves improvement in selectivity, high atom efficiency, elimination of hazardous reagents, and easy separation with recovery and reuse of reagents. 4*H*-Pyran derivatives represent an important class of compounds. They are often used in cosmetics and pigments, and utilized as potentially biodegradable agrochemicals. They also constitute a structural unit of many natural products and biologically interesting compounds which possess various pharmacological activities such as antiallergic and antitumor. In our efforts to develop a facile and greener one-pot multi-component reaction, the synthesis of 2-amino-4*H*-pyrans was achieved under solvent-free conditions using magnesium oxide as a catalyst in very good yields. The reaction catalyst, magnesium oxide, was reused and recycled without any loss of activity and product yields. All the synthesized compounds were screened for in vitro antibacterial activity. Some of these compounds showed complete inhibition of bacterial growth at 128 μg/ml or less, and the rest of the compounds exhibited incomplete inhibition at these concentrations.

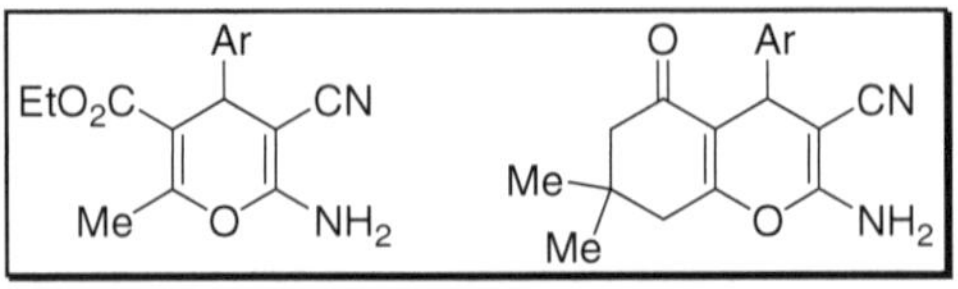

B.R. Vaddula (✉)
Chemistry Group, Birla Institute of Technology and Science, Pilani, Rajasthan-333 031, India
e-mail: buchi639@yahoo.co.in

Session 4
Novel Reagents for Multi-Component Reactions

Ketenimine Intermediate-Based Multi-component Reactions

Yanguang Wang

Ketenimines are a class of versatile and highly reactive intermediates that can participate in a variety of organic reactions, such as nucleophilic additions, radical additions, [2 + 2] and [2 + 4] cycloadditions, and sigmatropic rearrangements. In this presentation, we report on a series of multi-component reactions that involve a ketenimine intermediate. These reactions could furnish diverse heterocyclic compounds, including functionalized iminocoumarin, iminodihydroqunolines, iminothiochromens, pyrrolines, isoquinolines, pyridines, β-lactams, imino-1,2-dihydrocoumarins, and benzimidazoles.

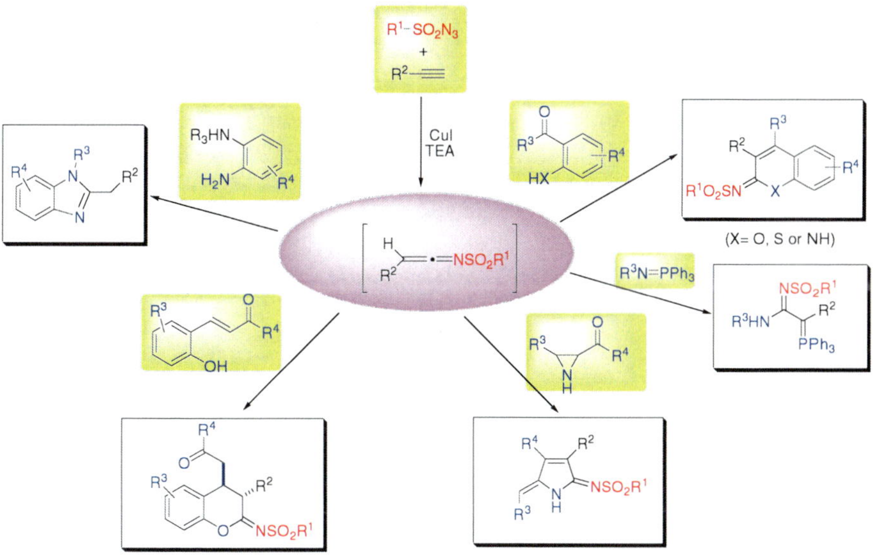

Y. Wang (✉)
Department of Chemistry, Zhejiang University, Hangzhou 310027, China
e-mail: orgwyg@zju.edu.cn

M.A. Mironov (ed.), *MCR 2009*, Advances in Experimental Medicine and Biology 699, 109
DOI 10.1007/978-1-4419-7270-5_4, © Springer Science+Business Media, LLC 2011

References

Cui SL, Lin XF, Wang YG. (2006) Novel and efficient synthesis of iminocoumarins via copper-catalyzed multicomponent reaction. Org Lett 8:4517–4520.

Cui SL, Wang J, Wang YG. (2008) Copper-catalyzed synthesis of 2-imino-1,2-dihydroquinolines and -thiochromenes. Tetrahedron 64:487–492.

Cui SL, Wang J, Wang YG. (2007) Copper-catalyzed multicomponent reaction: facile access to functionalized 5-arylidene-2-imino-3-pyrrolines. Org Lett 9:5023–5025.

Pluripotent Substrates in Diversity-Oriented Synthesis

Andrea Basso

Diversity-oriented synthesis represents a new methodological approach that in recent years has gained more and more importance in the field of organic synthesis. While traditional target-oriented synthesis is directed toward obtaining specific compounds, the goal of diversity-oriented synthesis is the efficient and automatizable preparation of collection of substances characterized by high diversity content, in terms of skeletons, appendages, and stereochemistries. One way to achieve this is to employ pluripotent substrates, which are molecules that can be synthetically elaborated, according to their functional groups, in many different ways, independent of each other, each way leading to a different molecular skeleton. Ideally the pluripotent substrate is not a single entity but a collection of molecules with a common skeleton, generated in a combinatorial fashion with the aid of multi-component reactions, already displaying all or most of the final decorative elements: This library of compounds is then further elaborated to generate *n* different combinatorial libraries, each with a different core structure. The ground-braking nature of this "star-burst" approach is that not only the various building blocks of the multi-component step but also the subsequent elaborations are introduced in a combinatorial fashion; thus, the final compounds, although deriving from a common substrate, are structurally different and therefore can cover a higher portion of the chemical space, compared to classic combinatorial libraries where all the members of one library are characterized by the same core.

The use of this approach to the design, synthesis, and biological applications of a diverse library of oxabicyclic derivatives, according to the figure illustrated below, will be discussed in detail.

A. Basso (✉)
Dipartimento di Chimica e Chimica Industriale, Università degli Studi di Genova,
Via Dodecaneso 31, 16146, Genova, Italy
e-mail: andreab@chimica.unige.it

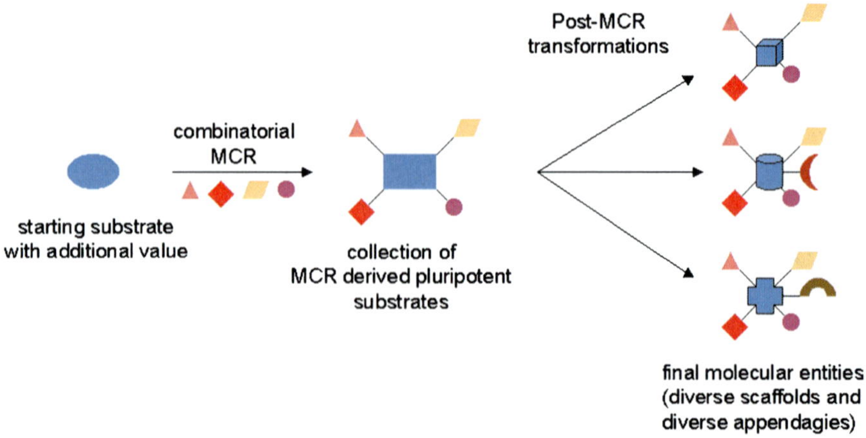

References

Basso A, Banfi L, Riva R. (2010) A marriage of convenience: combining the power of isocyanide-based multicomponent reactions with the versatility of (hetero)norbornene chemistry. Eur J Org Chem 2010(10):1831–1841.

Basso A, Banfi L, Guanti G, Riva R. (2010) Straightforward stereoselective synthesis of polyfunctionalized cyclohexenols using a multicomponent approach. Tetrahedron 66:2390–2397.

Di Micco S, Vitale R, Pellecchia M, Rega MF, Riva R, Basso A, Bifulco G. (2009) Identification of lead compounds as antagonists of protein Bcl-x(L) with a diversity-oriented multidisciplinary approach. J Med Chem 52:7856–7867.

Basso A, Banfi L, Galatini A, Guanti G, Rastrelli F, Riva R. (2009) Isocyanides and arylacetic acids: synthesis and reactivity of 3-aryl-2-acyloxyacrylamides, an example of serendipity-oriented synthesis. Org Lett 11:4068–4071.

Basso A, Banfi L, Guanti G, Riva R. (2009) A novel intramolecular Ugi reaction with 7-azabicyclo[2.2.1]heptane derivatives followed by post-condensation acylations: a new entry to azanorbornyl peptidomimetics. Org Biomol Chem 7:253–258.

Basso A, Banfi L, Riva R, Guanti G. (2006) Preparation of optically pure fused polycyclic scaffolds by Ugi reaction followed by olefin and enyne metathesis. Tetrahedron 62:8830–8837.

Basso A, Banfi L, Riva R, Guanti G. (2005) A novel highly selective chiral auxiliary for the asymmetric synthesis of L- and D-alpha-amino acid derivatives via a multicomponent Ugi reaction. J Org Chem 70:575–579.

Basso A, Banfi L, Riva R, Guanti G. (2004) U-4C-3CR versus U-5C-4CR and stereochemical outcomes using suitable bicyclic beta-amino acid derivatives as bifunctional components in the Ugi reaction. Tetrahedron Lett 45:587–590.

New Polyfunctional Reagents for Isocyanide-Based Multi-component Reactions

Valentine G. Nenajdenko and Anton V. Gulevich

Isocyanide-based multi-component reactions (IMCRs) are widely used in the fields of modern organic, combinatorial, and medicinal chemistry due to their exceptional synthetic efficiency and diversity-generating ability. The development of new starting materials for MCRs can increase the synthetic potential and open an approach to new types of products.

In the last decade, we have been involved in multi-component reactions of isocyanides, especially the synthesis of peptides and peptidomimetics by the Ugi or Passerini reactions. Synthetic and stereochemical aspects of IMCRs are also of interest to us.

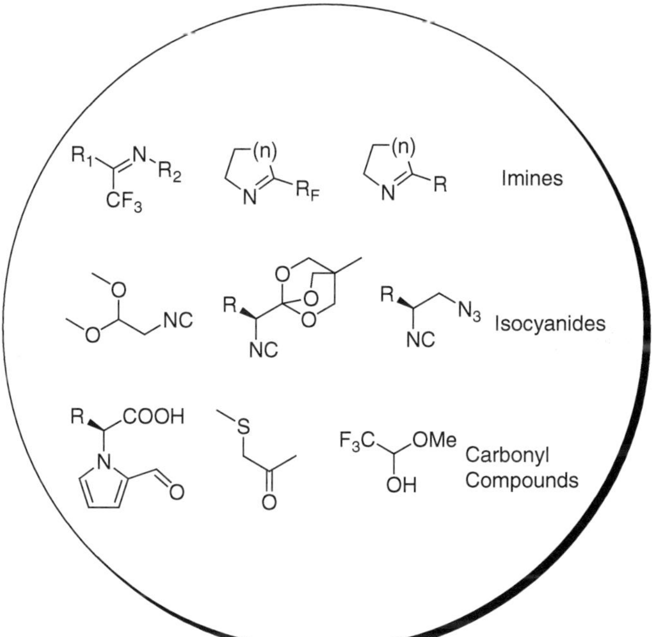

V.G. Nenajdenko (✉)
Moscow State University, Department of Chemistry, Leninskie Gory, Moscow, 119991, Russia
e-mail: nen@acylium.chem.msu.ru

References

Zhdanko AG, Nenajdenko VG. (2009) Nonracemizable isocyanoacetates for multicomponent reactions. J Org Chem 74:884.

Gulevich AV, Shpilevaya IV, Nenajdenko VG. (2009) The Passerini reaction with CF3-carbonyl compounds — multicomponent approach to trifluoromethyl depsipeptides. Eur J Org Chem 3801–3808.

Gulevich AV, Shevchenko NE, Balenkova ES, Röschenthaler G-V, Nenajdenko VG. (2008) The Ugi reaction with CF_3-carbonyl compounds: effective synthesis of α-trifluoromethyl amino acid derivatives. Tetrahedron 64:11706–11712.

Products of Ugi Reaction Among Acid Hydrazides, Aldehydes, and Isocyanides with Added Diversity

Mikhail Krasavin, Ekaterina Bushkova, and Vladislav Parchinsky

We revisited a long-known reaction of acid hydrazides as amine replacements in the classical Ugi reaction described by Ivar Ugi back in 1963 (Scheme 1).

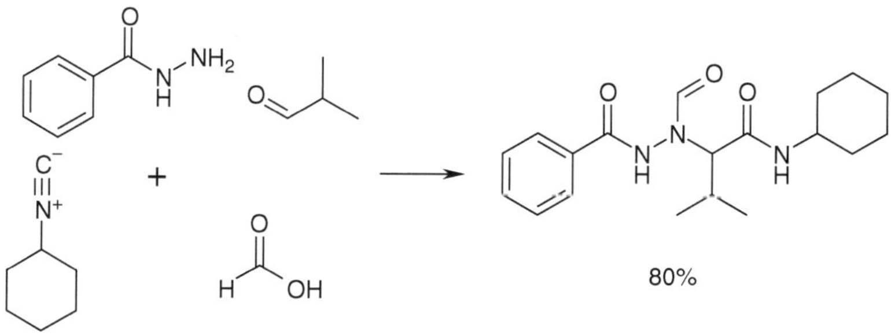

Scheme 1 Multi-component reaction toward a phenalenone derivative

We saw it as a drawback that this reaction leads to *N,N′*-bis-acyl hydrazines with limited room for further product diversification. In the course of studying this reaction, we found that when it is performed using trifluoroacetic acid as the carboxylate

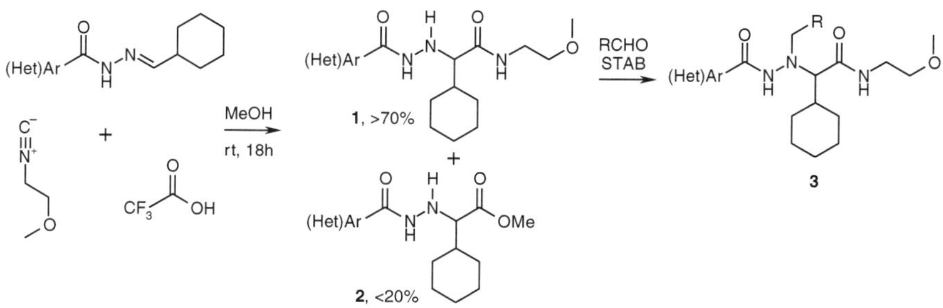

Scheme 2 Reductive amination of hydrazines

M. Krasavin (✉)
Chemical Diversity Research Institute, 2a Rabochaya St, Khimki, Moscow Reg, 141400, Russia
e-mail: myk@chemdiv.com

component, the products of the transformation are terminal *N*-monoacyl hydrazines **1** accompanied by esters **2** as minor byproducts. Furthermore, products **1** are suitable substrates for reductive amination, thus leading to *N*-acyl-*N'N'*-*bis*-alkyl hydrazines **3** (Scheme 2).

A discussion of the utility of the products **2** and **3** (HetAr = 4-pyridyl or pyrazinyl) as potential antitubercular agents (inspired by recent work of Dömling et al.) as well as our vision for possible chemistries based on the easy-to-prepare nucleophilic hydrazines **1** will be presented.

Reference

Dömling A et al. (2007) Novel anti-tuberculosis agents from MCR libraries. Bioorg Med Chem Lett 17:5483–5486.

Highly Functionalized Tricyclic *N*-Heterocycles by Coupling an Ugi MCR with a Tandem Pd(0)-Promoted S_N2'-Heck Reaction

Luca Banfi, Andrea Basso, Valentina Cerulli, Giuseppe Guanti, and Renata Riva

Multi-component reactions are very powerful methodologies for the convergent synthesis of very complex and unusual molecules provided that they are followed by suitable secondary transformations, and they can be regarded as an effective tribute to diversity-oriented synthesis.

In this context, we recently developed a very efficient synthesis of highly functionalized pyrrolidines by coupling a four-component Ugi reaction, involving

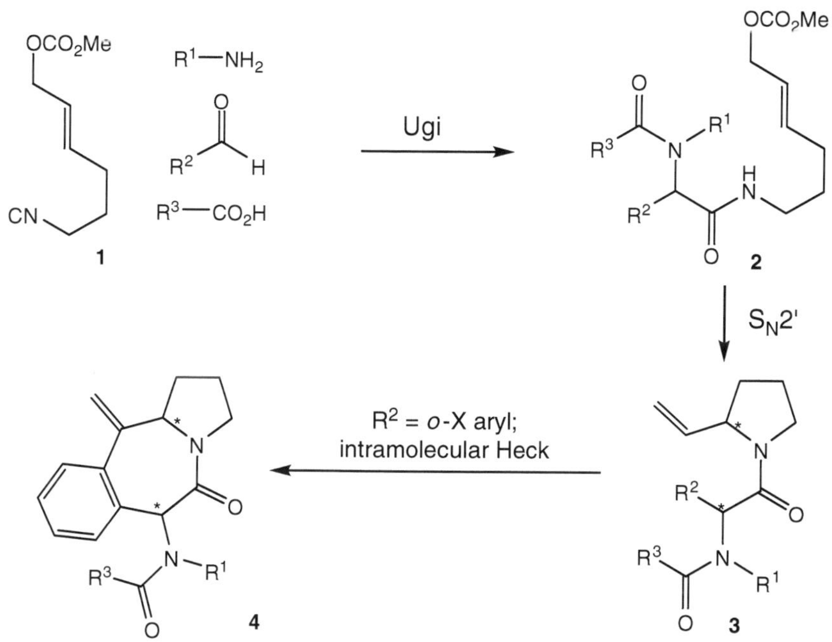

R. Riva (✉)
Università di Genova, Dipartimento di Chimica e Chimica Industriale,
Via Dodecaneso 31, 16146, Genova, Italy
e-mail: riva@chimica.unige.it

unsaturated isocyanide **1**, equipped with a carbonate, with a Pd(0)-mediated S_N2' cyclization promoted by the secondary amide to give compound **3**.

We further explored the scope of this protocol by adding an additional transformation, involving the terminal double bond of compound **3**, which is a Heck cyclization, which is possible if an aromatic aldehyde, with a suitable halide in the *ortho* position, is employed as carbonyl input in the MCR. During the optimization of our protocol, we found the best conditions for the direct transformation of acyclic derivative **2** into the complex structure **4**, without the need to isolate intermediate pyrrolidine **3**.

Reference

Banfi L, Basso A, Cerulli V, Guanti G, Riva R. (2008) Polyfunctionalized pyrrolidines by Ugi multicomponent reaction followed by palladium-mediated S_N2' cyclizations. J Org Chem 73:1608–1611.

Cascade Transformations of *N*-(1-Aryl-2-polychloroethyl)sulfonamides

Igor B. Rozentsveig, Gulnur N. Rozentsveig, Aleksandr V. Popov, Valeriy J. Serykh, and Galina G. Levkovskaya

A range of amidopolyhaloethylated aromatics (**1**) has been involved in organic synthesis as functionalized arenes containing synthetically attractive fragments: polyhalomethyl groups and protected amino groups. The synthetic significance of the compounds (**1**) has been demonstrated by the syntheses of biologically active amino acids, amidine systems, and chloroaziridines. This work deals with the cascade reactions of amides (**1**) in dipolar aprotic mediums under the action of *N*-, *S*-nucleophiles in the presence of inorganic bases.

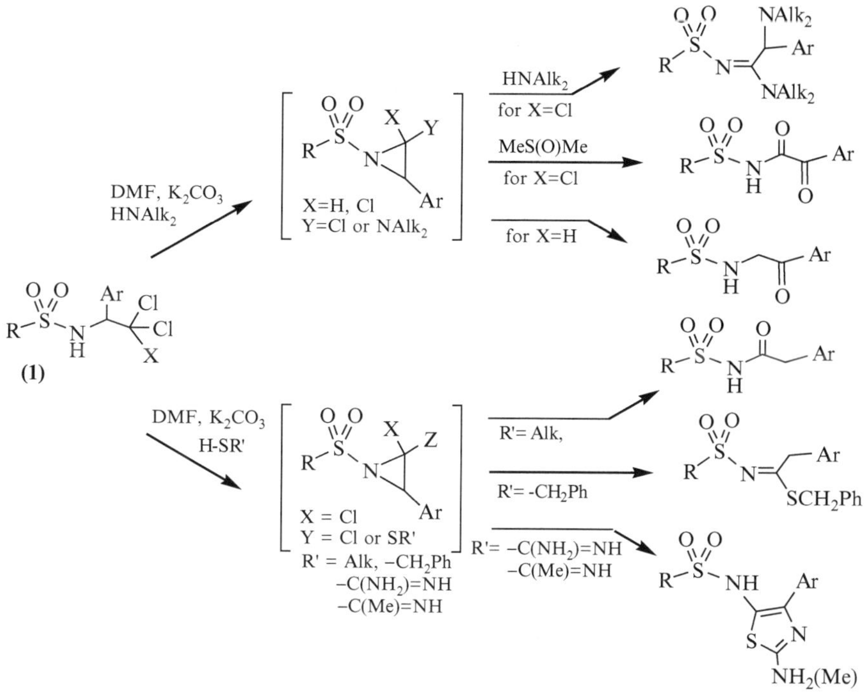

I.B. Rozentsveig (✉)
A.E. Favorsky Irkutsk Institute of Chemistry Siberian Division RAS,
1 Favorsky St, 664033, Irkutsk, Russia
e-mail: i_roz@irioch.irk.ru

The reactions appear to include the formation of aziridines that undergo the recyclization and subsequent transformations (aminolysis, hydrolysis or thiolysis, reduction or oxidation of chloromethylene groups) depending on the reaction conditions and reagents taken. The many-step, one-port sequence processes discovered are very efficient and convenient approaches to enamides, imides, amidinic systems, *N*-protected aminoketones, and thiazoles. Such cascade reactions hold much promise for obtaining functional amide derivatives from the compounds (**1**) and different nucleophiles as well.

Reference

Rozentsveig IB, Levkovskaya GG, Rozentsveig GN, Mirskova AN, Krivdin LB, Larina LI, Albanov AI. (2005) Amidine derivatives of α-arylglycines from *N*-(1-aryl-2,2,2-trichloroethyl) amides of arenesulfonic acids and secondary amines. Tetrahedron Lett 46:8889.

New Strategy for the Synthesis of Fluorinated Compounds: Multi-component Reactions Including Fluoro-containing Building Blocks

Song Cao, Li Shen, Nianjin Liu, Jingjing Wu, Lina Li, and Xuhong Qian

In recent years, multi-component reactions have become one of the most productive areas in organic synthesis and drug discovery. However, until now there were only a few reports on the use of fluoro-containing building blocks as one of the components in the multi-component reactions. Therefore, the strategy we proposed previously, of multi-component reactions including fluoro-containing building blocks (MCRs-FBB), will be a powerful tool for the one-pot construction of complex and diverse fluoro-containing molecules with high efficiency. In this presentation, we focus on and review the use of fluoro-containing building blocks as one of the components in multi-component reactions, and then introduce our research based on the MCRs-FBB strategy, using 5-trifluoromethyl-2,3-dichloropyridine, ethyl trifluoroacetoacetate, and a novel fluoro-containing building block, α,α-difluoro-β-amino amide, bearing the isocyanide functionality for the construction of fluoro-containing compounds.

Examples of application of the strategy of MCRs-FBB follow:

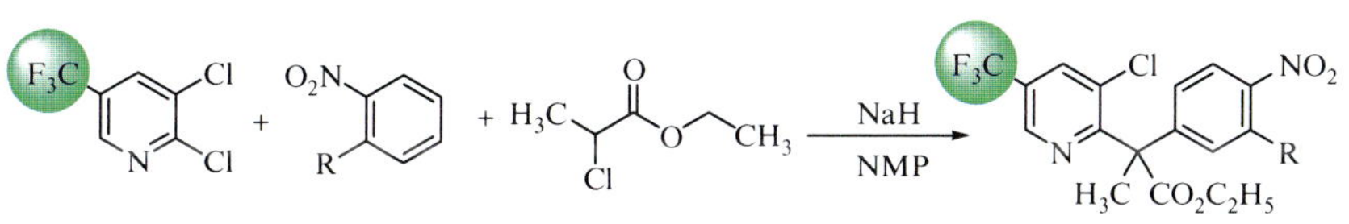

R = H, CH_3O, Cl, CH_3, $(CH_3)_2CHO$, $CH_2{=}CHCH_2O$

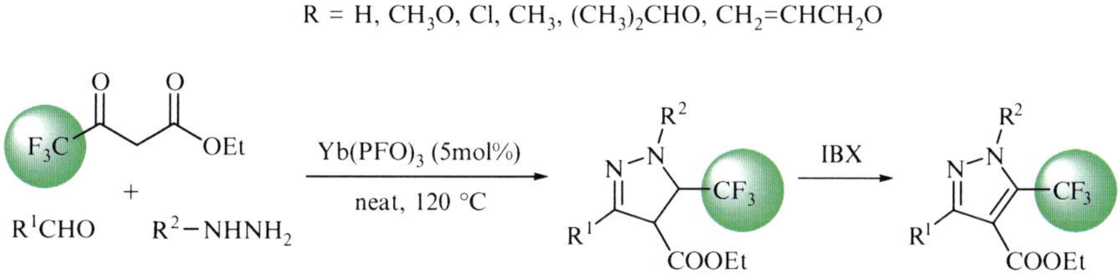

S. Cao (✉)
Shanghai Key Laboratory of Chemical Biology, Center of Fluorine Chemical Technology, School of Pharmacy, East China University of Science and Technology, Shanghai, 200237, China
e-mail: scao@ecust.edu.cn

References

Cao S, Wu J, Li L, Qian X, et al. (2008) One-pot three component synthesis of heteroarylated arylacetates via VNSAr-SN reaction using *N*-methylpyrrolidone as general solvent. Org Biomol Chem 6:1293–1297.

Shen L, Zhang J, Cao S, Qian X, et al. (2008) One-pot synthesis of trifluoromethyl-containing pyrazoles via sequential $Yb(PFO)_3$-catalyzed three-component reaction and IBX-mediated oxidation. Synlett 19:3058–3062.

A Multi-component Reaction Leading to a Phenalenone Derivative

Xiaopeng Chen, Hongbo Wang, Jinwu Feng, Yanguang Wang, and Ping Lu

A novel reaction catalyzed by $Pd(PPh_3)_2Cl_2$/CuI leading to a phenalenone derivative will be presented. This domino reaction eventually combines nine sequential steps in one pot, including Pd-catalyzed Sonogoshira coupling, Pd-catalyzed allylic oxidation, and Pd-catalyzed C–H activation. It is significantly worthwhile because all three types of C–H activations (Csp–H, Csp2–H, and Csp3–H) were concurrently involved in this transformation. Research work in this direction is in progress.

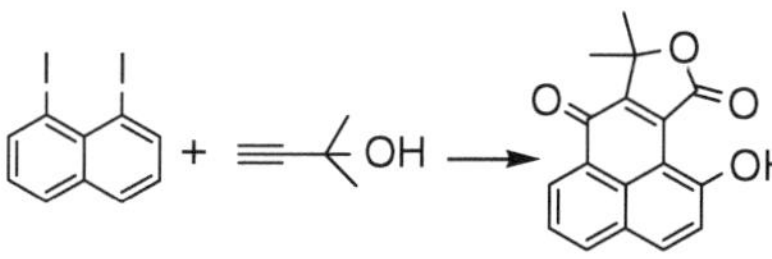

Reaction condition: $[Pd(PPh_3)_2Cl_2]$ (5 mol%), CuI (10 mol%), PPh_3 (10 mol%), trace of water and air, Et_3N, 100°C, 10 h, 64%.

Acknowledgment P. Lu thanks the National Nature Science Foundation of China (20674070). Contributions from Prof. Shengming Ma and Prof. Jinquan Yu in mechanism postulation and Jianming Gu and Xiurong Wu in structure elucidation are greatly appreciated.

References

Beccalli EM, Broggini G, Martinelli M, Sottocornola S. (2007) C–C, C–O, C–N bond formation on sp^2 carbon by Pd(II)-catalyzed reactions involving oxidant agents. Chem Rev 107:5318.

Zhao Z, Xu X, Wang H, Lu P, Yu G, Liu Y. (2008) Zigzag molecules from pyrene-modified carbazole oligomers: synthesis, characterization, and application in OLEDs. J Org Chem 73:594.

Zhao Z, Li J-H, Chen X, Lu P, Yang Y. (2008) Fluorescent conjugated dendrimers with fluorinated terminal groups: nanofiber formation and electroluminescence properties. Org Lett 10:3041.

Zhao Z, Li J-H, Lu P, Yang Y. (2007) Fluorescent, carrier-trapping dopants for highly efficient single-layer polyfluorene LEDs. Adv Funct Mater 17:2203.

P. Lu (✉)
Chemistry Department, Zhejiang University, 310027, Hangzhou, China
e-mail: pinglu@zju.edu.cn

Heteropoly Acids in One-Pot, Three-Component Synthesis of 4-Aminoquinazolines

Majid M. Heravi and Samaheh Sadjadi

Heteropoly acids (HPAs) are well-defined molecular clusters that are remarkable for their molecular and electronic structural diversity, and their significance is quite diverse in many areas, including catalysis, medicine, and materials science.

Quinazolines are classes of fused heterocycles that are of considerable interest because of the diverse range of their biological properties, for example, anticancer, diuretic, antiinflammatory, anticonvulsant, and antihypertensive activities.

In connection with our program of using HPAs in organic reactions, and extending our interest in multi-component reactions, and considering the low yields of some classical methods for the synthesis of 4-aminoquinazolines and lack of investigation toward the synthesis of these compounds in recent years, we investigated the synthesis of 4-aminoquinazolines from a one-pot reaction of 2-amino benzonitrile, acylchloride, and ammonium acetate in the presence of catalytic amounts of the Keggin type of HPAs, including $H_6[PMo_9V_3O_{40}]$, $H_5[PMo_{10}V_2O_{40}]$, $H_4[PMo_{11}VO_{40}]$, and $H_3[PMo_{12}O_{40}]$.

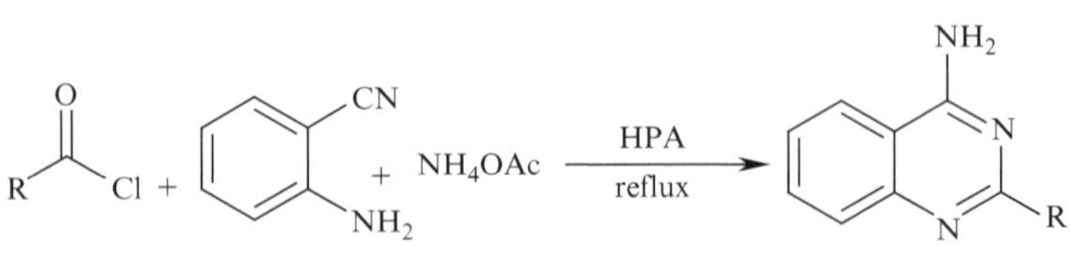

M.M. Heravi (✉)
Department of Chemistry, School of Sciences, Azzahra University, Vanak, Tehran, Iran
e-mail: heravi@gmail.com

References

Heravi MM, Alimadadi Jan B, Derikvand F, Bamoharram FF, Oskooie HA. (2008) Three component, one-pot synthesis of dihydropyrano[3,2-*c*]chromene derivatives in the presence of $H_6P_2W_{18}O_{62}$ $18H_2O$ as a green and recyclable catalyst. Catal Commun 10:272.

Misono M, Ono L, Koyano G, Aoshima A. (2000) Heteropolyacids. Versatile green catalysts usable in a variety of reaction media. Pure Appl Chem 72:1305.

Pope MT, Muller A. (1994) Polyoxometalates: from platonic solids to anti-retroviral activity. Dordrecht: Kluwer Academic Publishers.

Heravi MM, Sadjadi S, Oskooie HA, Hekmat Shoar R, Bamoharram FF. (2009) A direct oxidative route for the synthesis of pyrimidines using heteropolyacids. Tetrahedron Lett 50:662–666.

Heravi MM, Baghernejad B, Oskooie HA. (2009) A novel three-component reaction for the synthesis of *N*-cyclohexyl-3-aryl-quinoxaline-2-amines. Tetrahedron Lett 50:767–769.

Passerini Multi-component Reaction of Indane-1,2,3-trione, Isocyanides and 2-Furancarboxylic Acid

Ali Reza Kazemizadeh and Ali Ramazani

Isocyanide-based multi-component reactions have been around for 88 years, with the first described in 1921 and named after its founder, Passerini. In 1921, Passerini pioneered the use of isocyanides and successfully developed a three-component synthesis of α-acyloxycarboxamide by reaction of a carboxylic acid, an aldehyde, and an isocyanide. The Passerini reactions are beginning to find utility in the drug discovery process and in total syntheses of biologically relevant natural products. In this article, we report on the Passerini multi-component reaction among indane-1,2,3-trione (**1**), isocyanides (**2**), and 2-furancarboxylic acid (**3**) (Scheme 1).

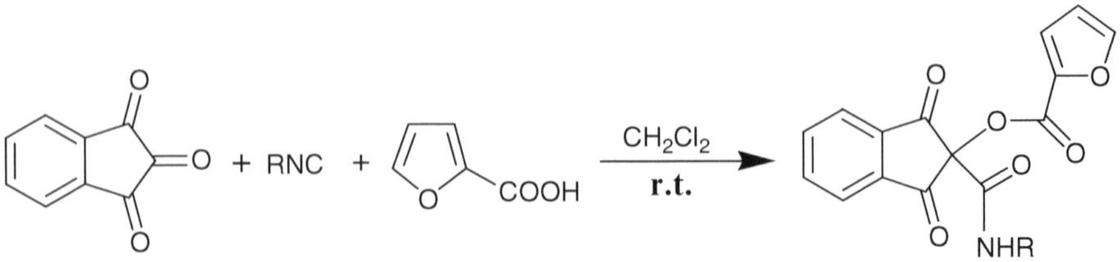

(1) **(2)** **(3)** **(4)**

Scheme 1 Indane-1,2,3-trione in the Passerini reaction

The structures of the products were deduced from their IR, ^1H NMR, and ^{13}C NMR spectra. The mass spectra of these compounds displayed molecular ion peaks at the appropriate *m/z* values.

References

Hulme C, Gore V. (2003) Multi-component reactions: emerging chemistry in drug discovery "from Xylocain to Crixivan". Curr Med Chem 10:51.

Ramazani A, Kazemizadeh AR, Ahmadi E, Noshiranzadeh N, Souldozi A. (2008) Synthesis and reactions of stabilized phosphorus ylides. Curr Org Chem 12:59.

A.R. Kazemizadeh (✉)
Department of Chemistry, Islamic Azad University-Zanjan Branch, 49195-467, Zanjan, Iran
e-mail: alirezakazemizadeh@yahoo.com

A Novel Route to the Synthesis of Fluoroalkylated Hexahydroimidazo[1,2-*a*] Pyridines

Yulia S. Kudyakova, Marina V. Goryaeva, Yanina V. Burgart, and Victor I. Saloutin

Interest in the synthesis of imidazo[1,2-*a*]pyridines is caused by their biological activity. Thus, their derivatives can be applied as calcium channel-blocking agents, kinase inhibitors, and substances with antiviral, antiulcer, and antibacterial properties.

A great number of such compounds' syntheses are illustrated in the literature. Two primary approaches should be noted: (1) condensation of 2-alkyl-imidazolines with various esters resulting in imidazopyridinones and (2) one-pot reaction of 2-aminopyridine with aromatic aldehyde and isocyanide under aqueous conditions.

In this work we have proposed a new synthesis to obtain fluoroalkylated imidazo[1,2-*a*]pyridines **4a–c** with good yields by the ternary condensation of 3-polyfluoroalkyl-3-oxoesters **1a–c** with acetone **2** and 1(*S*),2(*S*)-diphenyl-1,2-diaminoethane **3**.

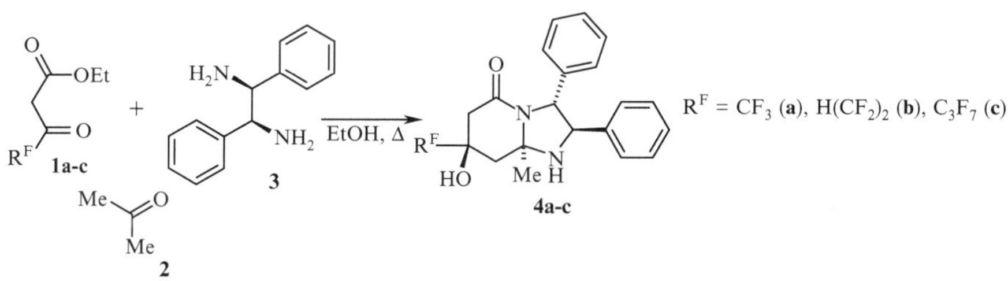

The structure of products **4a–c** was established on the basis of ^{1}H, ^{19}F NMR, IR spectroscopy, elemental analysis, and X-ray crystallography (see Fig. 1).

Y.S. Kudyakova (✉)
I.Ya. Postovsky Institute of Organic Synthesis, UB of RAS, 22/20 S. Kovalevskaya/Akademicheskaya St, 620041, Ekaterinburg, Russia
e-mail: kud@ios.uran.ru

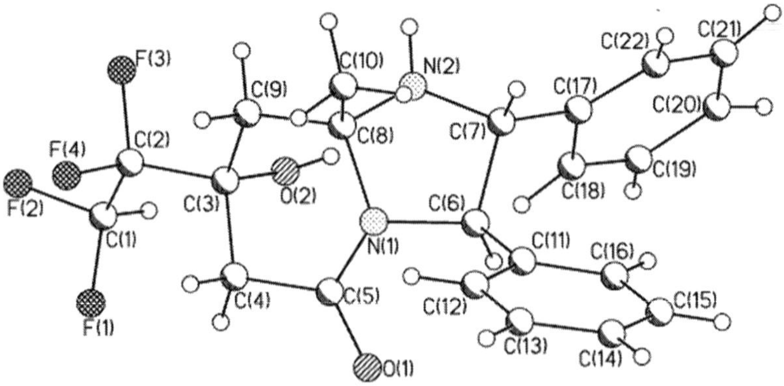

The molecular structure of 2,3-diphenyl-7-hydroxy-8a-methyl-7-(1,1,2,2-tetrafluoroethyl) hexahydroimidazo[1,2-*a*]pyridine-5-one 4b

Acknowledgment This study was supported by the Programme for the Support of Leading Scientific Schools (grant no. 3758.2008.3).

References

Townsend LB, Drach JC. (1997) Imidazo[1,2-a]pyridine C-nucleosides as antiviral agents. Int Pat WO 97/27205.

Gueiffier A, Mavel S, Lhassani M, Elhakmaoui A, Snoeck R, Andrei G, Chavignon O, Teulade J-C, Witvrouw M, Balzarini J, De Clercq E, Chapat J-P. (1998) Synthesis of imidazo[1,2-a] pyridines as antiviral agents. J Med Chem 41(25):5108–5112.

Jones RCF, Dimopoulos P, Coles SC, Light ME, Hursthouse MB. (2000) Synthesis of imidazo[1,2-a]pyridines from unactivated 2-alkyl-4,5-dihydroimidazoles through conjugate N-addition. J Chem Soc, Perkin Trans 1:2331–2342.

Katritzky AR, Xu Y-J, Tu H. (2003) Regiospecific synthesis of 3-substituted imidazo[1,2-a]pyridines, imidazo[1,2-a]pyrimidines, and imidazo[1,2-c]pyrimidine. J Org Chem 68:4935–4937.

Exploiting the Secondary Amines and the Glycolaldehyde Dimer in the Ugi Reaction: Formation of a Novel Passerini–Ugi Chimera Scaffold

Riccardo Mossetti, Tracey Pirali, and Gian Cesare Tron

The Ugi four-component reaction and the Passerini three-component reaction are the two most important and often-used isocyanide-mediated MCRs, and a large synthetic effort has been made since their discovery toward the full exploitation of these two venerable reactions. It does not escape the alerted chemist that all compounds which derive from an Ugi or a Passerini reaction display the same molecular scaffold, that is, NC(O)CNC(O) and NC(O)COC(O), respectively.

In this communication, we present our efforts to generate a novel molecular skeleton that can be formally viewed as a hybrid between Ugi and Passerini products, exploiting a secondary amine and the glicolaldehyde dimer in the Ugi reaction.

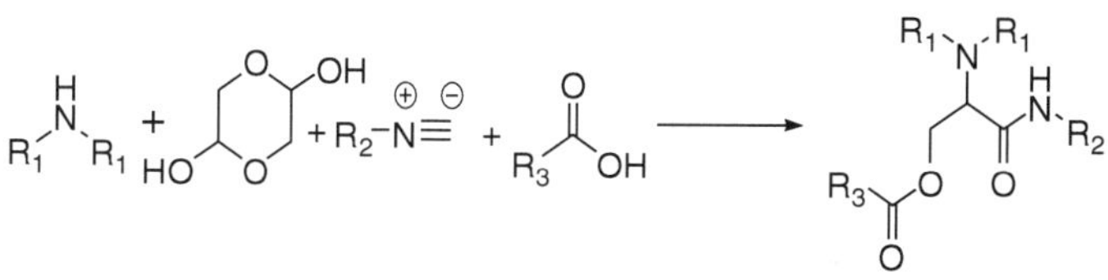

References

Marcaccini S, Torroba T. (2007) The use of the Ugi four-component condensation. Nat Protoc 2:632–639.

Riva R, Banfi L. (2005) Passerini reaction. Org React 65:1–140.

R. Mossetti (✉)
Università del Piemonte Orientale, DISCAFF, Via Bovio 6, 28100, Novara, Italy
e-mail: mossetti@pharm.unipmn.it

Three-Component Synthesis of 4′*H-spiro* [Adamantane-2,3′-isoqunolines]

Yulia S. Rozhkova, Olga A. Mayorova, Yuriy V. Shklyaev

A one-pot, three-component method was developed for the synthesis of 4′*H-spiro*[adamantane-2,3′-isoqunolines] from the reaction of activated arenes with 2-epoxymethyleneadamantane and nitriles in the presence of concentrated sulfuric acid. It was established that ortho-substituted arenes gave good yields of 4′*H-spiro*[adamantane-2,3′-isoqunolines] (**1**–**6**).

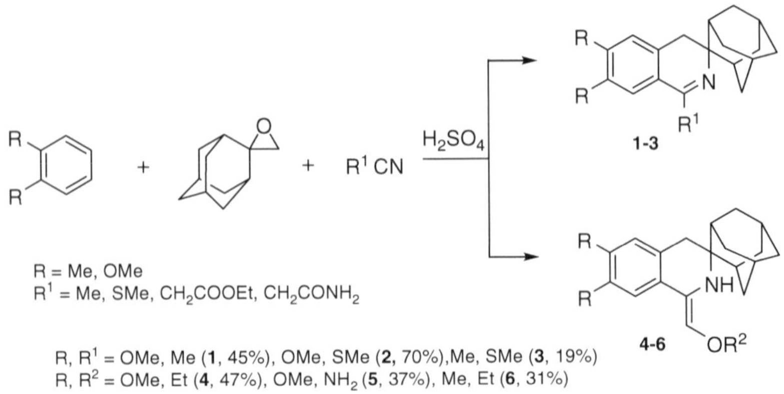

The position of substitutions on the aromatic ring has a defining effect on this reaction. It was shown that in the case of *para*-substituted arenes, 4′*H-spiro*[adamantane-2,3′-isoqunolines] (**7**, **8**, **10**, **11**) were obtained in disappointingly low yields. 4′*H*,5′*H-spiro*[adamantane-2,4′-oxazoles] (**9**, **12**) formed by the reaction of epoxide with nitrile were the main products in these reactions. This can be explained by the fact that methyl and methoxy-groups on *para*-substituted arenes create a steric hindrance for the acid-catalyzed alkylation arene by 2-epoxymethyleneadamantane.

Y.S. Rozhkova (✉)
Institute of Technical Chemistry of Urals Branch of Russian Academy of Sciences, Akademika Koroleva Street, 3, 614900, Perm, Russia
e-mail: rjs@mail.ru

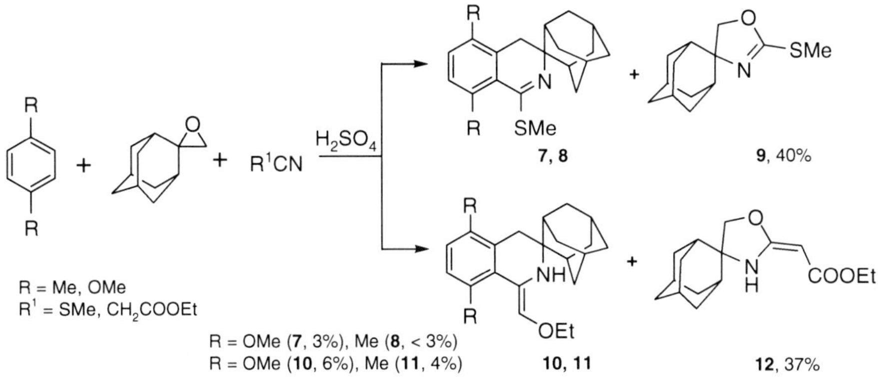

The structure of all the compounds **1–12** was confirmed by elemental analysis and IR and NMR [^{1}H, ^{13}C (DEPT)] spectral data.

Acknowledgment Our research has been performed with the financial support of the grant of the President of Russian Federation no. MK-2198.2008.3.

References

Aleksandrov BB, Gavrilov MS, Vakhrin MI, Shklyaev VS. (1985) Chem Heterocycl Comp 21(6):662.

Nifontov YV. (2001) Cand Sci (Chem.). Dissertation, Perm.

Glushkov VA, Shklyaev YV. (1998) Oxiranes in the Ritter reaction: synthesis of 6,7-(or 5,8-) dimethoxy-3,4-dihydroisoquinolines by a tandem alkylation—cyclization procedure. Mendeleev Commun 8:17.

Nonracemizable Isocyanoacetates for Multi-component Reactions

Alexander G. Zhdanko and Valentine G. Nenajdenko

α-Isocyano esters (isocyanoacetates) **1** have been used to date in the Ugi reaction (U-MCR) to introduce a *C*-terminal amino acid into the structure of desired peptide **2**. However, the following are disadvantages of the method: low reactivity of **1** (because of the COOR group influence), side reactions, and racemization (both because of the high CH-acidity of **1**). Therefore, we developed another type of α-amino acid isocyanide **3** for the synthesis of peptides by U-4CR.

Chiral ortho-esters of α-isocyano acids **3** were synthesized from commercially available Cbz-protected α-amino acids. These compounds are stable toward racemization in the Ugi 4CC, in contrast to known esters of α-isocyano acids. Applying them in the Ugi 4CC with subsequent deprotection gives access to dipeptides **4** with a preserved configuration at the *C*-terminal amino acid.

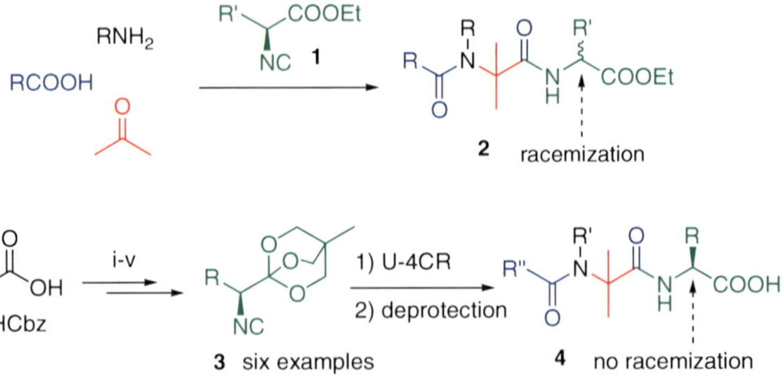

(i) 3-methyl-3-oxetanemethanol, DCC, DMAP, CH_2Cl_2, rt, >83%; **(ii)** $BF_3{\cdot}Et_2O$, CH_2Cl_2, rt, >95%; **(iii)** H_2/Pd-C, MeOH, rt, 99 %; **(iv)** HCOOEt, heat; or HCOOH, AcCl, Et_3N, CH_2Cl_2 80–92 %; **(v)** $POCl_3$, Et_3N, CH_2Cl_2, -20 °C, 81–90 %"

A.G. Zhdanko (✉)
Moscow State University, Department of Chemistry, Leninskie Gory, Moscow, 119991, Russia
e-mail: Vinceeero@gmail.com

References

Zhdanko AG, Gulevich AV, Nenajdenko VG. (2009) One-step synthesis of N-acetylcysteine and glutathione derivatives using the Ugi reaction. Tetrahedron 65:4692–4702.

Zhdanko AG, Nenajdenko VG. (2009) Nonracemizable isocyanoacetates for multicomponent reactions. J Org Chem 74:884–887.

Short Communication: Synthesis

Three-Component Synthesis of 2-Aza-spiro[4.5] decane Derivatives

Olga G. Stryapunina, Irina V. Plekhanova, Vladimir A. Glushkov, and Yurii V. Shklyaev

Three-component electrophilic domino-condensation of 1,2,3-, 1,2,4-, and 1,3,5-trimethoxy-substituted benzenes, alpha-branched aldehydes, and nitriles in the presence of concentrated sulfuric acid was studied. The reaction afforded substituted 2-aza-spiro[4.5]deca-6,9-dienes (**1**), 1-alkoxycarbonylmethylidene-6,10-dimethoxy-8-[(α-cyano-α-alkoxycarbonyl)-methylidene-3-R^4,3-R^5-2-aza-spiro[4.5] deca-6,9-dienes (**2**), or 1,2-dimethoxy-3-oxo-14-azadispiro[5.1.5.2]pentadeca-1,4,14-trienes (**3**), depending on the character of the arene and aldehyde. In the case of aromatic nitriles, along with 1-aryl-3,3-dimethyl-2-aza-spiro[4.5]deca-1,6,9-trienes (**4**), the considerable quantities of normal Ritter amides (**5**) were also obtained.

V.A. Glushkov (✉)
Institute of Technical Chemistry, Acad. Korolev Street, 3, 614013, Perm, Russia
e-mail: ushka@newmail.ru

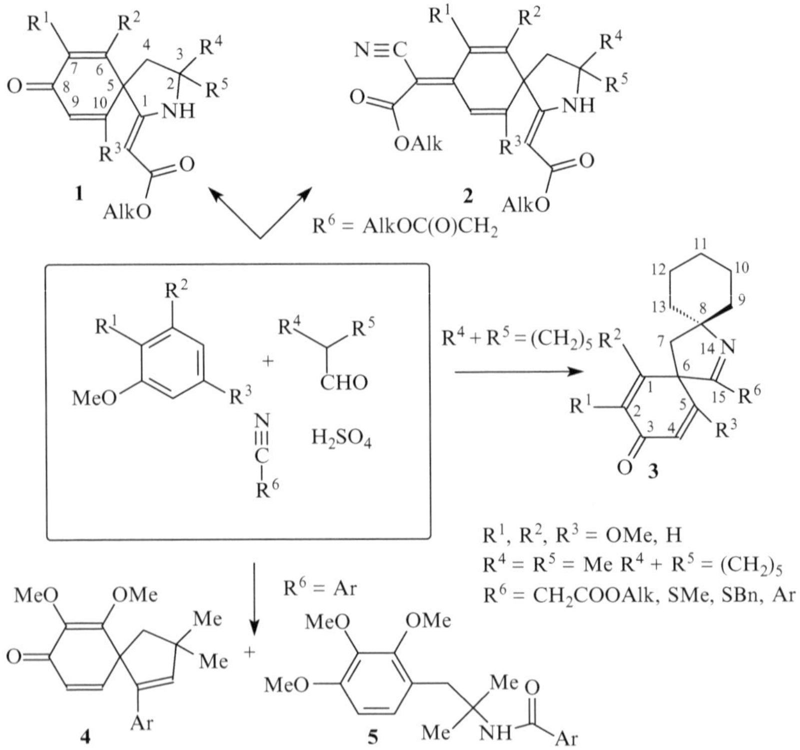

The structure of all compounds was confirmed by elemental analysis, NMR ^{1}H and ^{13}C spectra, and, for some compounds, single-crystal X-ray diffraction.

Acknowledgment This work was supported by the Presidium of the Russian Academy of Sciences (coordinator: academician V. A. Tartakovsky).

Session 5
Design of Multi-Component Reactions

Isocyanide-Based Multi-component Reactions: New Variations

Jieping Zhu

Multi-component reactions (MCRs) have now been well established as a powerful synthetic tool for creating molecular complexity and diversity and are undoubtedly well suited for the drug discovery program. Another potential that has probably received less attention among synthetic chemists is the opportunity offered by MCRs for the development of new fundamentally important transformations (reactions). Indeed, although an MCR is composed of a series of known bimolecular reactions, the overall transformation could be novel. Consequently, it provides chemists the opportunities to uncover transformations that were otherwise difficult to realize. In this talk, we will present our recent work in this field, including: (1) the oxidative homologation of aldehydes to amides, (2) the oxidative coupling of aldehydes and isocyanides to α-ketoamides, (3) oxidative isocyanide-based MCRs, and (4) the enantioselective Passerini reaction.

References

Grassot JM, Masson G, Zhu J. (2008) Synthesis of α-ketoamides by a molecular-sieves-promoted formal oxidative coupling of aliphatic aldehydes with isocyanides. Angew Chem Int Ed 47: 947–950.

Ngouansavanh T, Zhu J. (2007) IBX-mediated oxidative Ugi-type multicomponent reactions: application to the N and C1 functionalization of tetrahydroisoquinoline. Angew Chem Int Ed 46:5775.

Yue T, Wang M-X, Wang DX, Zhu J. (2008) Asymmetric synthesis of 5-(1-hydroxyalkyl)tetrazoles by catalytic enantioselective Passerini-type reactions. Angew Chem Int Ed 47:9454–9457.

J. Zhu (✉)
Institut de Chimie des Substances Naturelles, CNRS, 91198, Gif-sur-Yvette Cedex, France
e-mail: zhu@icsn.cnrs-gif.fr

M.A. Mironov (ed.), *MCR 2009*, Advances in Experimental Medicine and Biology 699,
DOI 10.1007/978-1-4419-7270-5_5,

From Ugi–Smiles Couplings to Nef Reaction of Isocyanides

Laurent El Kaïm

In the last 20 years, most synthetic work related to the Ugi reaction has involved the formation of heterocycles by postcondensation processes. Besides several studies on the radical cyclization of classical Ugi adducts, our efforts on this reaction have mostly focused on the design of efficient surrogates for carboxylic acid. In 2005 we proposed the use of electron-deficient phenols via a Smiles rearrangement as the key step in the process. Since then, we have extended the scope of these reactions to the use of various hydroxy- and mercapto-substituted heterocycles (Scheme 1).

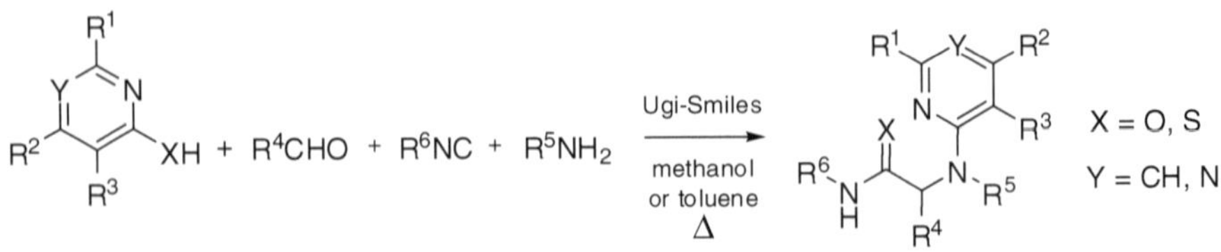

Scheme 1 Ugi–Smiles reactions

The synthetic power of these "Ugi–Smiles" couplings has been further demonstrated by several postcondensations giving access to azepine, quinoxaline, indole, benzimidazole or benzotriazole scaffolds.

In the second part of this presentation, new synthetic sequences involving a Nef coupling of isocyanides with acyl chlorides will be presented. The multi-component nature of the process is associated with the trapping of the Nef adducts by various nucleophiles, such as tetrazoles (Scheme 2) or phosphites (Scheme 3).

L.E. Kaïm (✉)
Unité Chimie et Procédés, Ecole Nationale Supérieure de Techniques Avancées,
32 Bd Victor, Paris, 75015, France
e-mail: laurent.elkaim@ensta.fr

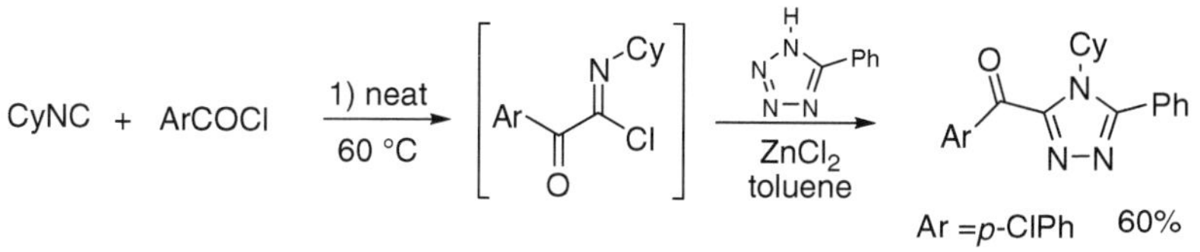

Scheme 2 New Nef–Huisgen sequence to triazole scaffolds

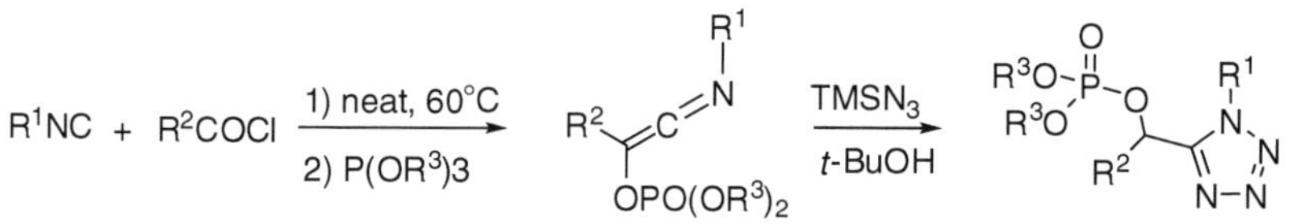

Scheme 3 Ketamine formation

References

El Kaim L, Gizolme M, Grimaud L, Oble J. (2007) Smiles rearrangements in Ugi- and Passerini-type couplings: new multicomponent access to *O*- and *N*-arylamides. J Org Chem 72:5835–5838.

Oble J, El Kaïm L, Gizzi M, Grimaud L. (2007) Ugi-Smiles access to quinoxaline derivative. Heterocycles 73:503–517.

El Kaim L, Gizzi M, Grimaud L. (2008) New MCR–Heck–isomerization cascade toward indoles. Org Lett 10:3417–3419.

El Kaim L, Gizolme M, Grimaud L, Oble J. (2007) Smiles rearrangements in Ugi- and Passerini-type couplings: new multicomponent access to *O*- and *N*-arylamides. J Org Chem 72:4169–4180.

The Importance of Being a Secondary Amine: From the N-Split Ugi Reaction of the Reshuffling of the Ugi and Passerini Products

Gian Cesare Tron

The most popular isocyanide-based multi-component reaction (IMCR) is the Ugi four-component reaction that produces linear α-acylaminoamides. Usually, the Ugi reaction uses primary amines, which play a dual role in the Ugi mechanism, being simultaneously alkylated and acylated, causing the loss of both hydrogen atoms. When a secondary amine is used, at least *three different* synthetic pathways can take place. This happens because the secondary amine can no longer undergo the transacylation reaction (Mumm rearrangement).

With the aim of discovering new IMCRs, we thought to use a secondary amine, in order to avoid the Mumm rearrangement, *freezing* the imino-anhydride intermediate for the appropriate time to allow it to be intramolecularly intercepted by other nucleophiles (e.g., nitrogen and oxygen).

Here we present our findings regarding the discovery of the *N*-split Ugi reaction and the synthesis of hybrid Passerini–Ugi scaffolds and their applications in medicinal chemistry.

G.C. Tron (✉)
Università del Piemonte Orientale, DISCAFF, Via Bovio 6, 28100, Novara, Italy
e-mail: tron@pharm.unipmn.it

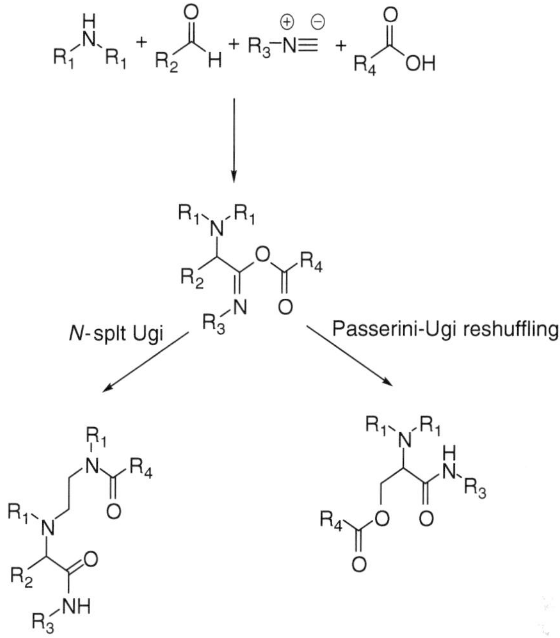

References

Giovenzana GB, Tron GC, Di Paola S, Menegotto IG, Pirali T. (2006) A mimicry of primary amines by bis-secondary diamines as components in the Ugi four-component reaction. Angew Chem Int Ed 45:1099–1102.

Pirali T, Tron GC, Zhu JP. (2006) One-pot synthesis of macrocycles by a tandem three-component reaction and intramolecular [3+2] cycloaddition. Org Lett 8:4145–4148.

Rebuilding a Forgotten MCR: Studies on the Interaction of Isocyanides with Alkylboranes and Aldehydes

Rodolfo Lavilla

The participation of boron reagents in MCRs is synthetically useful but somewhat restricted to boronic acids, and most of the research done in this area deals with the Petasis reaction. We considered that electrophilic boron reagents may interact with isocyanides to yield intermediate species suitable for multi-component (MC) processes. A literature survey was conducted; we found that early in the 1960s, G. Hesse described a striking MCR involving tributylborane, aryl isocyanides, and aromatic aldehydes. This work has not received attention, as any reference citing the original article was found in the databases.

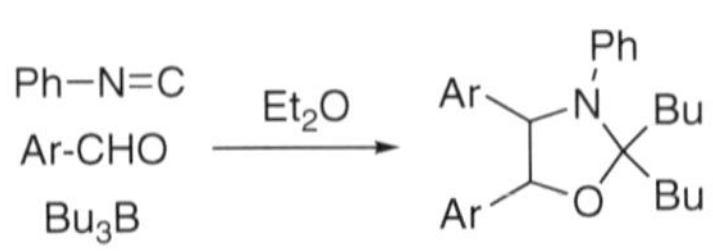

The scope of the reaction has been analyzed, and a preliminary mechanistic hypothesis has led to reaction variants allowing the inclusion of a fourth component. The description of these new processes together with some synthetic applications of potential interest in diversity-oriented synthesis will be discussed.

References

Kielland N, Catti F, Bello D, Isambert N, Soteras I, Luque FJ, Lavilla R. (2010) Boron-based dipolar multicomponent reactions: simple generation of substituted aziridines, oxazolidines and pyrrolidines. Chem Eur J 16(26):7904–7915.

Witte H, Gulden W, Hesse G. (1968) Liebigs Ann Chem 716:1–10.

Hesse G, Witte H, Gulden W (1965) Angew chem 77:591.

Hesse G, Witte H, Gulden W. (1965) Preparation of 1,3-oxazolidines from isonitriles and aldehydes by a multicomponent reaction. Angew Chem 77:591; Angew Chem Int Ed Engl 4:569.

R. Lavilla (✉)
Institute for Research in Biomedicine – Barcelona, C/ Baldiri Reixac 10-12, 08028, Barcelona, Spain
e-mail: rlavilla@pcb.ub.es

Diversity-Oriented Synthesis of Various Benzo-Fused Oxaza Heterocycles by Coupling the Ugi Reaction with Nucleophilic Substitutions

Luca Banfi, Andrea Basso, Valentina Cerulli, Giuseppe Guanti, Paulina Lecinska, and Renata Riva

The Ugi multi-component reaction is a very powerful tool in diversity-oriented synthesis, allowing the simultaneous introduction of four diversity inputs in just one step with a tremendous increase in structural complexity. However, the classic version of this reaction does not lead to "drug-like" heterocyclic systems and is not well suited for the exploration of scaffold diversity.

These drawbacks may be overcome by placing one or two additional functional groups in the starting components and exploiting them for a postcondensation cyclization. Multiple scaffolds can be obtained by simply varying the length or nature of the spacers bearing the additional functions or placing them in different components. We are presently demonstrating this possibility by preparing a series of different benzo-fused oxaza heterocycles, always using the same additional functionalities (a phenol and an alcohol) and the same postcondensation reaction (an aliphatic nucleophilic substitution). Selected examples, leading to scaffolds **1–3**, will be reported in this communication. The application of scaffold **1** to the combinatorial synthesis of potential α-helix mimics will be also described.

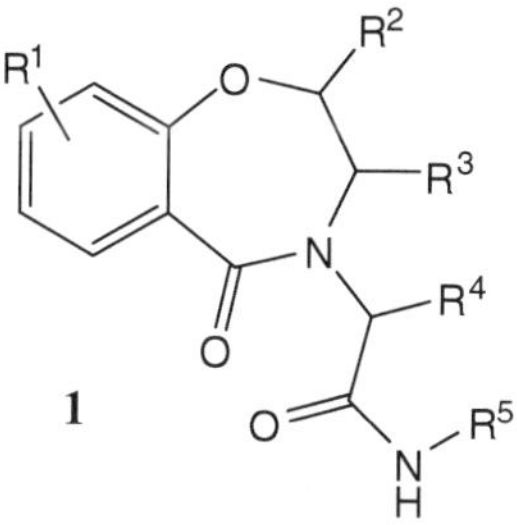

1

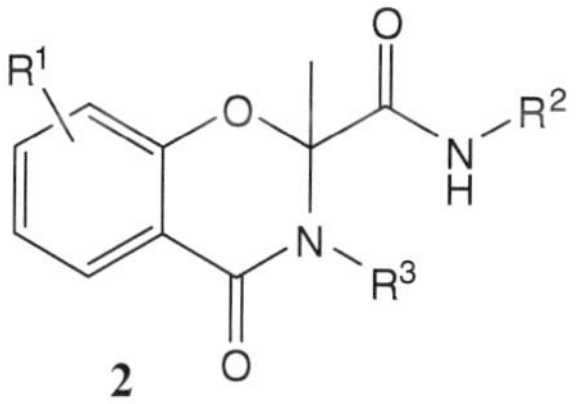

2

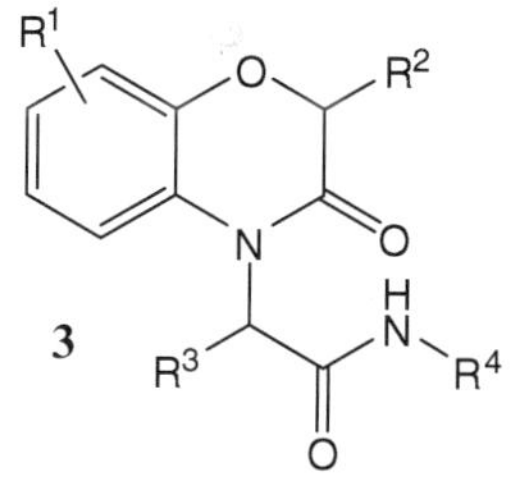

3

L. Banfi (✉)
Università di Genova, Dipartimento di Chimica e Chimica Industriale,
Via Dodecaneso 31, 16146, Genova, Italy
e-mail: banfi@chimica.unige.it

References

Banfi L, Basso A, Guanti G, Lecinska P, Riva R, Rocca V. (2007) Multicomponent synthesis of novel 2- and 3-substituted dihydrobenzo<1,4>oxazepinones and tetrahydro-benzo<1,4>oxazepinones and their conformational analysis. Heterocycles 73:699–728.

Banfi L, Basso A, Guanti G, Lecinska P, Riva R. (2006) Multicomponent synthesis of dihydrobenzoxazepinones by coupling Ugi and Mitsunobu reactions. Org Biomol Chem 4:4236–4240.

Banfi L, Basso A, Guanti G, Lecinska P, Riva R. (2008) Multicomponent synthesis of benzoxazinones via tandem Ugi/Mitsunobu reactions: an unexpected *cine*-substitution. Mol Diversity 12:187–190.

New Multi-Component Reactions Based on Electrophilic Activation of Azines

M. J. Arévalo, N. Kielland, C. Masdeu, M. Miguel, N. Isambert, and R. Lavilla

In recent years, the development of new multi-component reactions based on the activation of azines has proved to be an effective approach toward the synthesis of new heterocyclic compounds. In 2004 we described an efficient Ugi–Reissert reaction involving the interaction of isoquinolines, chloroformates, and isocyanides in which the activated azine is carbamoylated at the α-position by the isocyanide to give dihydroazines such as **4**. Looking for new activating agents, we explored the interaction of isoquinolines and pyridines with TFAA and isocyanides. These compounds interact in a different manner, yielding mesoionic acid fluorides with an imidazo[1,2-*a*]azine core, in a new domino process. This multi-component reaction has a general character, tolerating a wide range of substitution patterns on each component, and displays an unprecedented arrangement of reactivity pathways (Scheme 1).

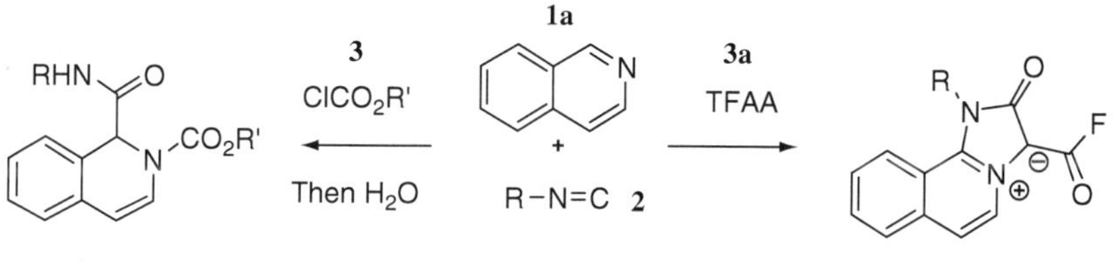

Scheme 1 Three-component reaction with isocyanides

N. Kielland (✉)
Institute for Research in Biomedicine (IRB Barcelona), C/Baldiri Reixac 10, 08028, Barcelona, Spain
e-mail: nicola.kielland@irbbarcelona.org

The protocol allows the incorporation of a fourth synthetic input through the reaction of a suitable nucleophile (alcohols, thiols, and amines) with the acid fluoride moiety. In the same way, **5** has been coupled with biomacromolecules. The obtained derivatives were suitable to enter into cells, and a preliminary study resulted in low toxicity levels in concentration levels up to 40 μmol. New synthetic and biological aspects of this chemistry will be discussed.

Reference

Díaz JL, Miguel M, Lavilla R, (2004) *N*-acylazinium salts: a new source of iminium ions for Ugi-type processes. J Org Chem 69:3550–3553.

Multi-Component Synthesis of New Heterocyclic Compounds from Propynals

Alevtina S. Medvedeva, Vladimir V. Novokshonov, Irina A. Novokshonova, Maria M. Demina, and Tatyana V. Kon'kova

New regio- and stereospecific microwave-assisted three-component reactions of trimethylsilyl-germylpropynals with trimethylsilylazide and primary amines occur in good to excellent yields of *E*-5-R-imino-[(4-trialkylsilyl(germyl)]-1-1,2,3-triazoles (**2**) or 1-1,2,3-triazolo-1′,3′-thiazolidines (**1**). Microwave irradiation allows considerable acceleration of reactions (domestic MW oven, 700 W; 60% power, 60 min, 90–95°; or 10–15 days, 25°).

Biginelli reactions of the α-element containing propynals or γ-hydroxypropynals with ethyl acetoacetate and urea catalyzed by HCl (5 mol%) or PPE give new acetylenic 3,4-dihydropyrimidinones (**3**) in 40–81% yield (MeOH or THF, reflux, 24 h). Microwave-mediated synthesis of the compound (**3**, R = Et_3Ge) was carried out on PPE/SiO_2 without solvent within 5 min (420 W, 65–70°) in 90% yield. The combination of the Biginelli reaction and 1,3-dipolar cycloaddition of azide ion in a four-component process leads to new triazolopyrimidinones (**4**) in good yields.

A.S. Medvedeva (✉)
A.E. Favorsky Irkutsk Institute of Chemistry, Siberian Branch of Russian Academy of Sciences, Favorsky Street, 1, 664033, Irkutsk, Russia
e-mail: amedved@irioch.irk.ru

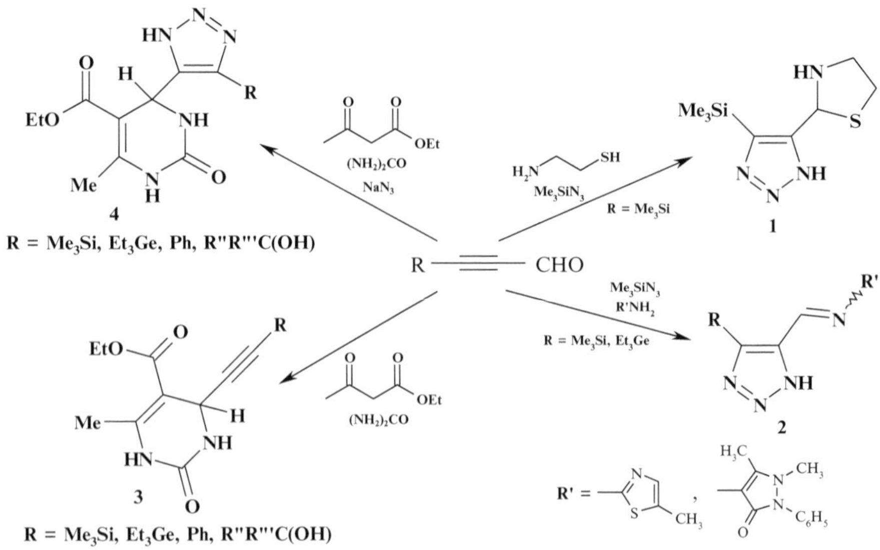

The high efficiency of multi-component approaches for the synthesis of new functionalized 3,4-dihydropyrimidinones and 1*H*-1,2,3-triazoles from ambident-substituted propynals has been demonstrated.

References

Novokshonov VV, Novokshonova IA, Ushakov IA, Medvedeva AS. (2006) Element-substituted propynals in the Biginelli reaction. Chem Heterocycl Comp 42:1942–1943.

Medvedeva AS, Mareev AV, Demina MM. (2008) Silicon- and germanium-substituted propynals as ambident electrophiles for the design of new heterocycles and polyfunctional acetylenes. Russ Chem Bull Int Ed 57:929–949.

Synthesis of Partially Hydrogenated Heterocycles: 3,4-Dihydroisoqunolines, Spiropyrrolines, Polyhydroindols

Yurii V. Shklyaev, Yulia S. Rozhkova, Tatiana S. Vshivkova, Olga G. Stryapunina, Vladimir A. Glushkov, and Anastasia V. Kharitonova

This report summarizes results of the three-component interaction among α-branched aldehydes (oxirans), nitriles, and activated arenes. These reactions lead to the formation of different heterocycles such as 3,4-dihydroisoqunolines, spiropyrrolines, polyhydroindols, and neospirans.

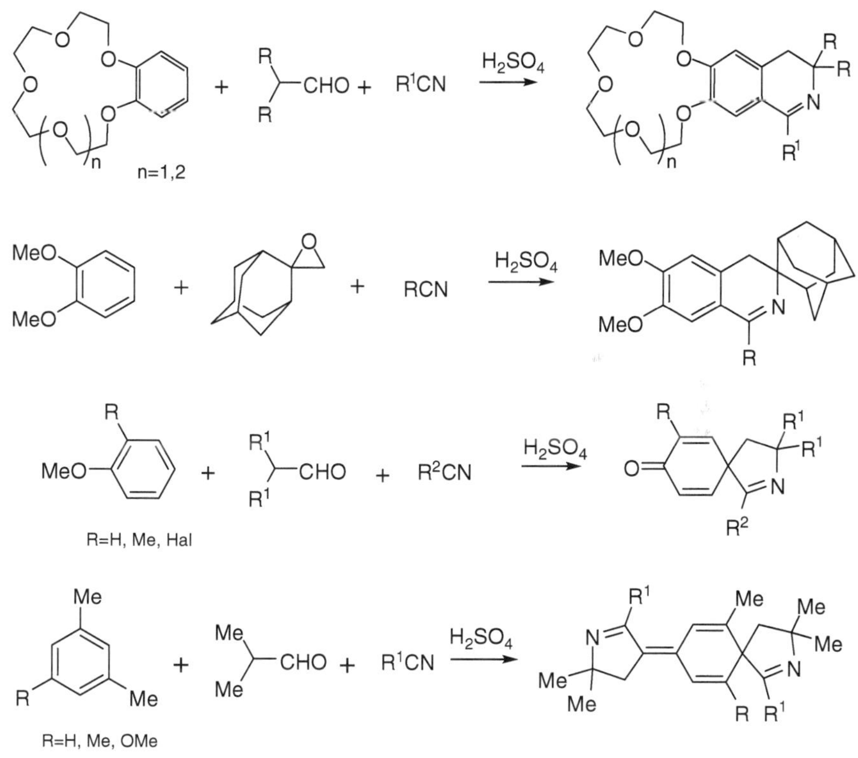

Y.V. Shklyaev (✉)
Institute of Technical Chemistry of Urals Branch of Russian Academy of Sciences, Akademika Koroleva Street, 3, 614900, Perm, Russia
e-mail: yushka@newmail.ru

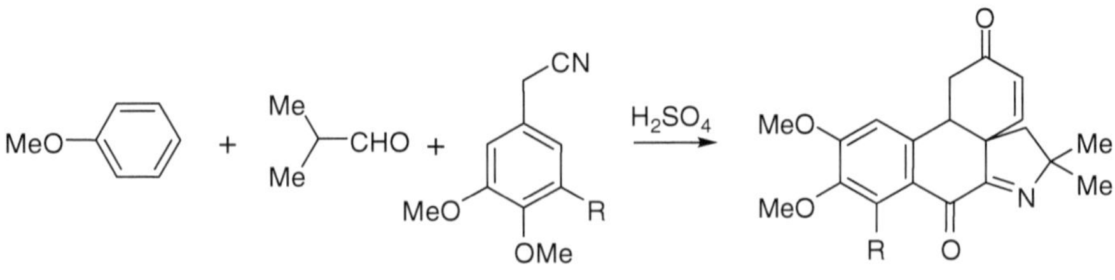

R=H, OMe

Acknowledgment Our research has been executed with the financial support of grant no. RFFI 07-03-0001, the program of Presidium of the Russian Academy of Sciences, on the theme "Development of Methods of Obtaining of Chemical Substances and Creation of New Materials."

References

Shklyaev YV, Eltsov MA, Maiorova OA. (2010) Three-component condensation of o- and p-methoxytoluenes with isobutyraldehyde and -substituted benzyl cyanides. Synthetic approach to analogs of natural alkaloids. Russ J Org Chem 46(1):113–116.

Rozhkova YS, Maiorova OA, Shklyaev YV. (2009) Synthesis of 4H-spiro[adamantane-2,3-isoquinoline] derivatives. Russ J Org Chem 45(12):1874–1876.

Glushkov VA, Stryapunina OG, Gorbunov AA, Maiorova OA, Slepukhin PA, Ryabukhina SA, Khorosheva EV, Sokol VI, Shklyaev YV. (2010) Synthesis of 1-substituted 2-azaspiro[4.5]deca-6,9-diene-8-ones and 2-azaspiro[4.5]deca-1,6,9-triene-8-ones by a three-component condensation of 1,2,3-, 1,2,4- or 1,3,5-trimethoxybenzene with isobutyric aldehyde and nitriles. Tetrahedron 66:721–729.

Shklyaev YV, Gilev MY, Maiorova OA. (2009) Retropinacol rearrangement in the synthesis of, 3,4-trimethyl-3,4-dihydroisoquinoline derivatives. Russ J Org Chem 45(12):1843–1846.

Development of Bi- and Multi-component Reactions for the Synthesis of Some Fused Hexahydropyrimines

Alexander S. Fisyuk, Aleksey Y. Mukanov, and Nicolay V. Poendaev

We have developed some bi- and multi-component reactions for the synthesis of 1,2,3,6,7,11b-hexahydro-4*H*-pyrimido[6,1-*a*]isoquinolin-4-ones (thiones) **1** and 2,3,6,7,12,12b-hexahydropyrimido[1′,6′:1,2]pyrido[3,4-*b*]indol-4(1*H*)-ones (thiones) **2**.

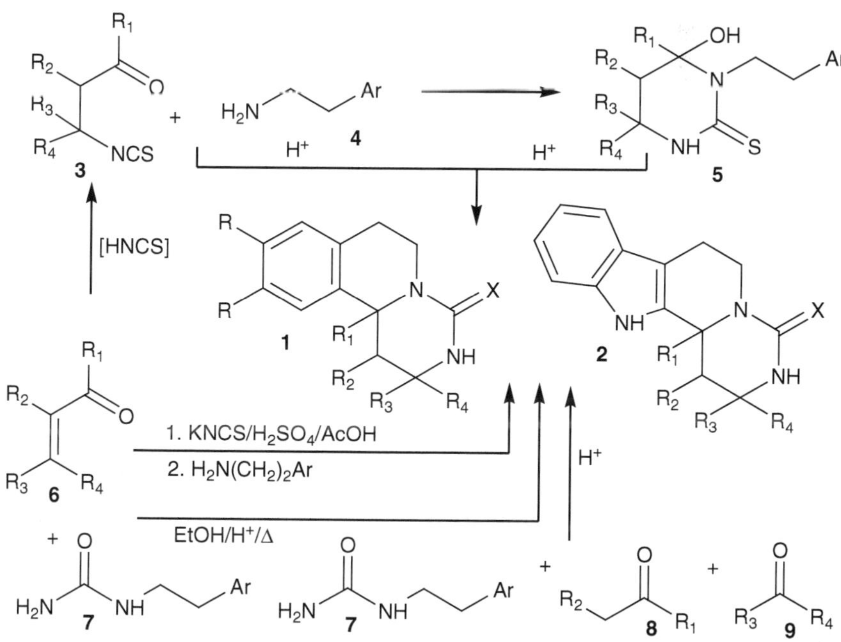

R_{1-4} = H,Alk; R= H,OH,OMe; X= O,S; Ar = C_6H_3-3,4-R_2; 3-Indolyl

A.S. Fisyuk (✉)
Chemistry Department, Omsk State University, pr. Mira, 55f644077, Omsk, Russia
e-mail: fis@orgchem.univer.omsk.su

The compounds **1**, **2** (X = S) were obtained by condensation of 1,3-isothiocyanatocarbonyl compounds with phenylethylamines or tryptamine **4** in acid media and one-port synthesis from α,β-unsaturated carbonyl compounds, in situ-generated thiocyanic acid and amines **4**. Compounds **1**, **2** (X = O) were obtained by condensation of α, β-unsaturated carbonyl compounds with *N*-ethylphenyl ureas **7** and *N*-ethylphenyl ureas **7** with carbonyl compounds **8**, **9**.

We have found the effects of substitutions R, R_1–R_4 on the conditions, yields, and products structures of the above-mentioned reactions.

References

Fisyuk AS, Poendaev NV. (2001) Synthesis and reactions of 5,6-dihydropyridin-2(1H)-ones and – thiones (review). Targets Heterocycl Syst 5:271.

Levi MD, Fisyuk AS, Demadrille R, Markevich E, Gofer Y, Aurbachand D, Pron A. (2006) Unusually high stability of a poly(alkylquaterthiophene-alt-oxadiazole) conjugated copolymer in its n and p-doped states. Chem Commun (31):3299–3301.

Giraud M, Andreiadis ES, Fisyuk AS, Demadrille R, Pécaut J, Imbert D, Mazzanti M. (2008) Efficient sensitization of lanthanide luminescence by tetrazole-based polydentate ligands. Inorg Chem 47(10):3952–3954.

Isocyanide-Based Multi-component Reactions with CF3-Carbonyl Compounds

Anton V. Gulevich and Valentine G. Nenajdenko

The development of new types of components for MCRs can increase the synthetic potential of the reaction. We focus our attention on polyfluorocarbonyl compounds possessing high carbonyl activity and other unique properties.

CF3-carbonyl compounds **1** and **2** have been investigated in the Passerini reaction. Based on the reaction, we developed a new approach to CF_3-depsipeptides – interesting candidates for cancer treatment. Unfortunately, CF_3-carbonyl compounds **1** or **2** are not effective in the Ugi reaction. However, reaction with preformed imines **3** opens a new route to α-Tfm-amino acid derivatives and CF_3-peptides. Easily available cyclic imines **4** were found to be very effective precursors for the synthesis of α-Tfm-substituted proline and homoproline derivatives.

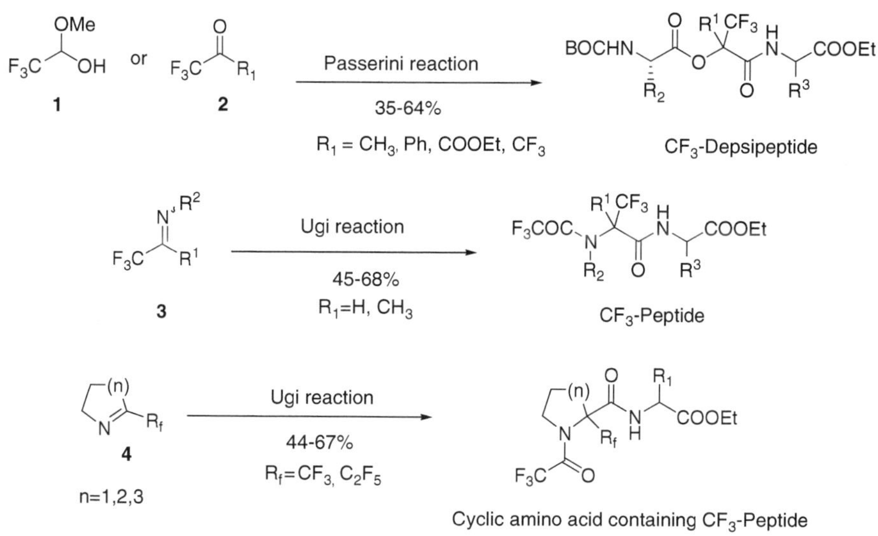

A.V. Gulevich (✉)
Moscow State University, Department of Chemistry, Leninskie Gory, Moscow, 119991, Russia
e-mail: anton.gulevich@synth.chem.msu.ru

References

Gulevich AV, Shevchenko NE, Balenkova ES, Röschenthaler G-V, Nenajdenko VG. (2008) The Ugi reaction with CF_3-carbonyl compounds: effective synthesis of α-trifluoromethyl amino acid derivatives. Tetrahedron 64:11706–11712.

Gulevich AV, Shevchenko NE, Balenkova ES, Röschenthaler G-V, Nenajdenko VG. (2009) Efficient multicomponent synthesis of α-trifluoromethyl proline, homoproline, and azepan carboxylic acid dipeptides. Synlett 3:403–406.

The Investigation of Three-Component Reaction Based on Isocyanides with Olefins and (Thio)phenols

Maria N. Ivantsova, Maria I. Tokareva, Maxim A. Mironov, and Vladimir S. Mokrushin

In the earlier publications our research group described the original search methodology of novel multi-component reactions. This methodology has allowed us to synthesize new and different heterocyclic compounds, such as imidazoisoquinoline, phthalazine, propionamide, dihydrothiophene-2,5-diimine, and 3-iminoindole derivatives. Recently the three-component reaction of isocyanides **2** with olefins **1** and 4-nitrophenol **3** has been discovered (Scheme 1). The products **4** and **5** were isolated with good yields (45–85%).

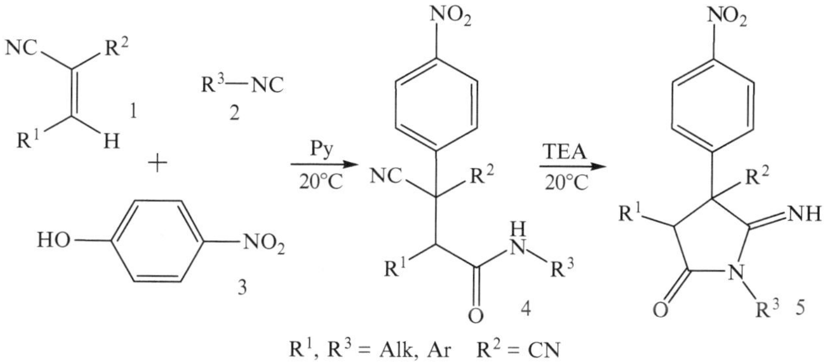

Scheme 1 Three-component reaction with isocyanides

The investigation of this reaction was continued. We have varied two parameters: Olefin and phenol and analogous products were synthesized with good yields (40–75%).

M.N. Ivantsova (✉)
SEI HPE (USTU-UPI named after the first President of Russia, Boris N. Yeltsin), Mira Street, 19, 620002, Ekaterinburg, Russia
e-mail: mivantsova@mail.ru

Furthermore, we have investigated the influence of thiophenol derivatives on the reaction course. We have isolated two type of heterocyclic compounds **7** and **8** in dependence on the structure of thiophenols **6** (Scheme 2).

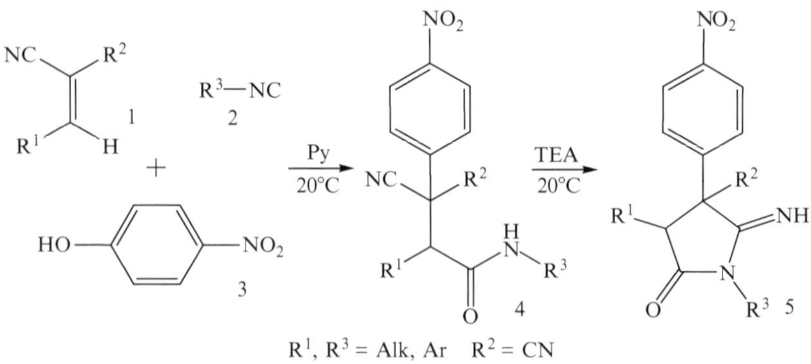

Scheme 2 Three-component reaction with isocyanides

The structure of all products is based on the detailed analysis of ^{1}H, ^{13}C spectra and X-rays.

References

Mironov MA, Ivantsova MN, Mokrushin VS. (2006) A novel isocyanide-based multicomponent reaction: an easy access to substituted propionamides and succinimides. Synlett 4:615–618.

Mironov MA, Mokrushin VS, Maltsev SS. (2003) New method for the combinatorial search of multi component reactions. Synlett 943–945.

Mironov MA, Tokareva MI, Mokrushin VS. (2007) Synthesis of 3-iminoindole derivatives by the reactions of aromatic isocyanides with *N,N*-dialkylbenzylamines. Mendeleev Commun 17:354–356.

Mironov MA, Ivantsova MN, Tokareva MI, Mokrushin VS. (2007) Novel synthesis of dihydrothiophene-2, 5-diimine derivatives by the three-component reaction of isocyanides with enamines and arylisothiocyanates. Heterocycles 73:567–579.

Improving the Isocyanide-Based Nef Synthesis: Reaction Between Acyl Chlorides and α-Isocyanoacetamides

Tracey Pirali, Gian Cesare Tron, and Jieping Zhu

The reaction between acyl chlorides and isocyanides to produce α-ketoimidoyl chlorides is commonly referred to as the Nef reaction. Over the last decades, these intermediates have been used as synthetic precursors thanks to their ability to give the electrophilic acylnitrilium species. In particular, they can be converted upon hydrolysis into α-ketoamides, a privileged scaffold in medicinal chemistry, which can also be found in a number of biologically important natural products. However, the condensation of isocyanides with acyl chlorides is known to be problematic and lacks in generality, and few protocols have been reported so far.

In this communication, we present our findings about the unprecedented reaction between acyl chlorides and α-isocyanoacetamides to produce 2-keto-5-aminooxazole and, after hydrolysis, α-aminodiketoamides.

Further modifications of the oxazole intermediates will also be presented.

T. Pirali (✉)
Università degli Studi del Piemonte Orientale, DiSCAFF,
Via Bovio 6, 28100, Novara, Italy
e-mail: pirali@pharm.unipmn.it

Nef Reaction

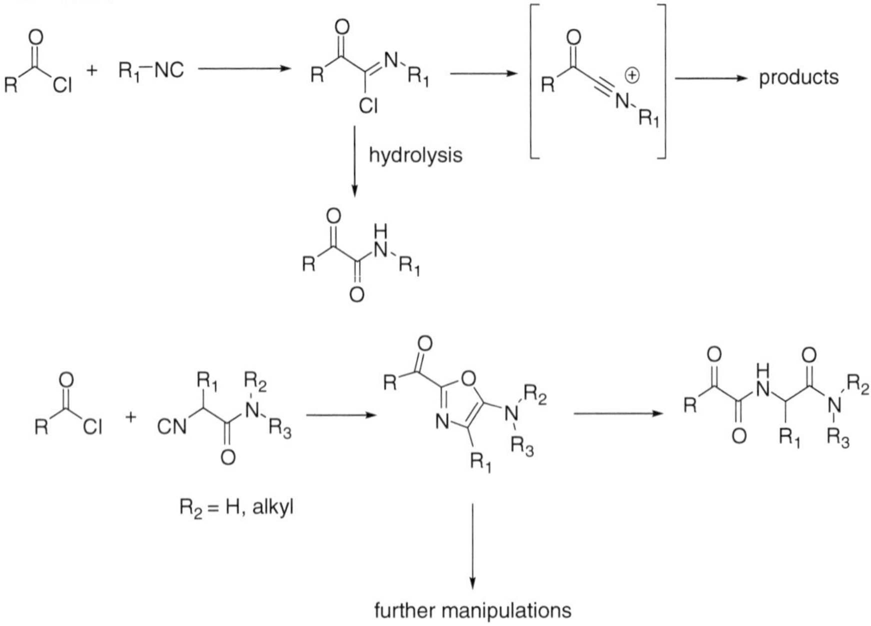

References

Chen JJ, Deshpande SV. (2003) Rapid synthesis of α-ketoamides using microwave irradiation–simultaneous cooling method. Tetrahedron Lett 44:8873.

Livinghouse T. (1999) *C*-Acylnitrilium ion initiated cyclizations in heterocycle synthesis. Tetrahedron 55:9947.

Schreiber SL. (1991) Chemistry and biology of the immunophilins and their immunosuppressive ligands. Science 251:283.

Cascade Reactions of N-(1-Aryl-2-Polychloroethyl)-Sulfonamides with Mercaptoethanol

Igor B. Rozentsveig, Aleksandr V. Popov, Galina G. Levkovskaya, Kirill A. Chernyshev, and Leonid B. Krivdin

Efficient and facile synthetic methods have been previously developed for the synthesis of amidoalkylated aromatic derivatives of type (**1**). They are based on easily available reagents: dichloroamides, polyhaloethenes, and aromatics. Different transformations of sulfonamides (**1**) into functionalized sulfonamide species are being studied extensively at the present time. This work deals with the cascade reactions of trichloroethylamides and dichloroethylamides of type (**1**) with mercaptoethanol in the presence of inorganic bases in dipolar aprotic media.

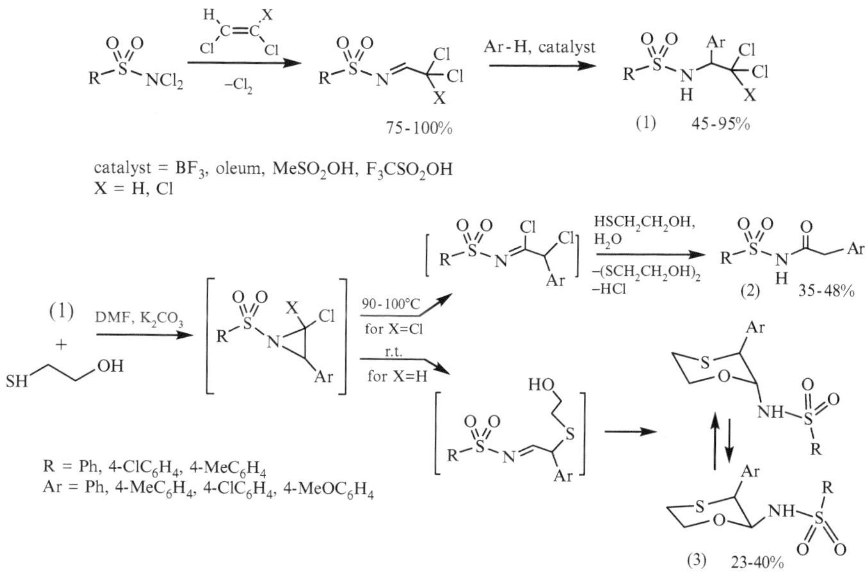

I.B. Rozentsveig (✉)
A.E. Favorsky Irkutsk Institute of Chemistry, Siberian Division of the Russian Academy of Sciences, 1 Favorsky Street, 664033, Irkutsk, Russia
e-mail: i_roz@irioch.irk.ru

It was found that trichloroethylamides of arenesulfonic acids with mercaptoethanol under heating in DMF in the presence of potassium or sodium carbonates afford *N*-sulfonylamides of arylacetic acids (**2**) as final products. The formation of the compounds (**2**) probably includes the cyclization to aziridines, which undergo the recyclization and subsequent transformations: isomerization, hydrolysis of the chloroimidoyl fragment, and reduction of carbon in the α-position. The reaction does not take place without heating. Dichloroethylamides react with mercaptoethanol at room temperature and produce *N*-(3-aryl-1,4-oxathian-2-yl)arenesulfonamides (**3**), which exist in the form of different conformers.

References

Levkovskaya GG, Drozdova TI, Rozentsveig IB, Mirskova AN. (1999) Usp Khim 68:638.
Rozentsveig IB, Evstaf'eva IT, Levkovskaya GG, Mirskova AN. (2000) Russ J Org Chem 36:847.
Rozentsveig IB, Levkovskaya GG, Albanov AI, Mirskova AN. (2000) Russ J Org Chem 36:671.

A New Type of Cascade Reactions, Proceeding in the Presence of Diazocarbonyl Compounds

Yury V. Tomilov and Dmitry N. Platonov

A new methodology for the synthesis of functionally substituted pyrazoles, piridazines, and their particularly hydrogenated analogs on the basis of the cascade reaction of α-diazocarbonyl compounds with pyridinium and triphenylphosphonium ylides was elaborated. The process proceeds through the addition of diazoester or diazoketone to the ylide carbon atom by the terminal nitrogen atom with elimination of the ylide-forming molecule followed by the addition of an additional one or two (in the case of $Py^{+}–C^{-}HCOR$) ylide fragments to intermediate azines. One of the determining stages is the interaction of diazocompound with the ylide molecule; diazoalkanes and substituted compounds of the N_2CRCO_2Me, $N_2C(CO_2Me)_2$ type do not react similarly. Subsequent addition of ylide fragments to reactive azines occurs until fully substituted azine is formed. The latter depends on the reaction conditions, the nature of the ylide-forming molecule, and substituents in both reagent forms of substituted azaheterocycles, which are of interest for new biologically active compounds' design.

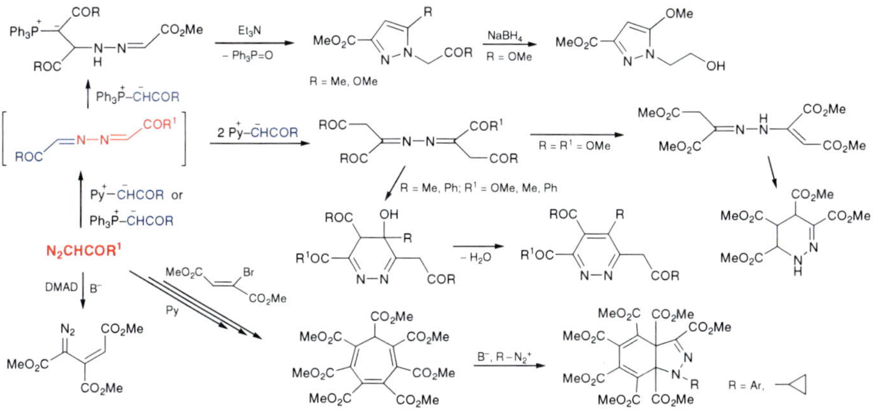

Y.V. Tomilov (✉)
N. D. Zelinsky Institute of Organic Chemistry, Russian Academy of Sciences,
Leninsky prosp., 47, 119991, Moscow, Russia
e-mail: tom@ioc.ac.ru

New directions for the interaction of diazoesters with electron-deficient unsaturated compounds in the presence of bases were developed as well. Thus, the interaction of methyl diazoacetate with dimethyldibromosuccunate or dimethylbromofumarate in pyridine gives heptamethoxycarbonylcyclohepta-1,3,5-triene as a result of successive cascade reactions accompanying Py·HBr elimination and cyclization in a final stage. The compound is capable of forming stable cycloheptatrienyl anions reacting with electrophilic reagents, among them aryl- and cyclopropyldiazonium.

The nucleophilic addition of diazoester to acetylene dicarboxylate in the presence of DABCO leads to the formation of a new diazocompound – diazoaconitic acid triester.

References

Tomilov YV, Platonov DN, Dorokhov DV, Nefedov OM. (2007) A new method for the synthesis of azaheterocycles based on cascade reactions of nitrogen- and phosphorus-containing ylides with methyl diazoacetate. Tetrahedron Lett 48:883.

Tomilov YV, Platonov DN, Salikov RF, Okonnishnikova GP. (2008) Synthesis and properties of stable 1,2,3,4,5,6,7-heptamethoxycarbonylcyclohepta-2,4,6-trien-1-yl potassium and its reactions with electrophilic reagents. Tetrahedron 64:10201.

Multistep Process of Nucleophilic Vinylic Substitution of Captodative Halogeno(Carbonyl)Alkenes

Alexander Y. Rulev and Igor A. Ushakov

Captodative carbonyl-containing halogeno- and aminoalkenes have been the object of long-lasting interest because of their direct involvement in the chemistry of various bioactive compounds. They are versatile building blocks in the synthesis of various important carbo- and heterocycles, as well as analogs of natural substances. One of the very direct routes to the captodative carbonyl-bearing aminoalkenes relies on the S_NV in reaction of the corresponding halogenoalkenes with *N*-nucleophiles.

From a mechanistic point of view, the formation of aminoalkenes **2** is not a one-step process. The key steps of these domino transformations seem to be as follows: (1) aza–Michael addition of amine to the double bond of the halogenoalkenes **1**,leading to an intermediate α-bromo-β-aminoderivative; (2) nucleophilic substitution of the halogen at the *sp*^{3}-carbon atom; (3) β-elimination of an amine molecule forming the captodative carbonyl-containing aminoalkene. In the case of the strong electron-withdrawing group $C(O)CF_3$, the following hydroxyalkylation reaction of the intermediates **2** leads easily to the indenols **3**.

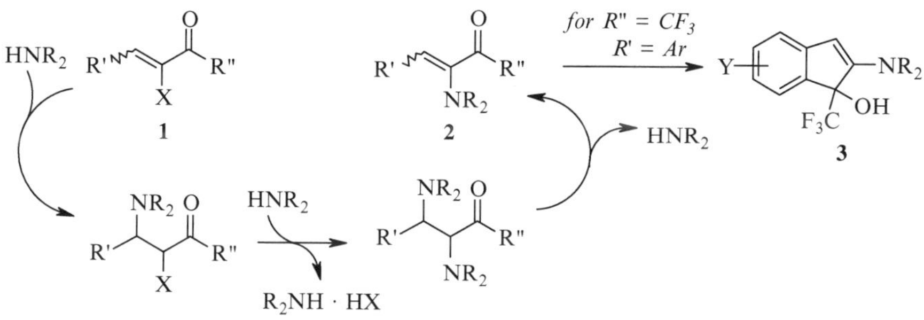

Chemical and spectroscopic evidence for the postulated scheme is discussed.

A.Y. Rulev (✉)
A. E. Favorsky Institute of Chemistry, Siberian Branch of the Russian Academy of Sciences, 1 Favorsky Street, Irkutsk, 664033, Russia
e-mail: rulev@irioch.irk.ru

Acknowledgment This work was supported by the Russian Foundation for Basic Research (grant no. 08-03-00067-a).

Reference

Rulev AY, Ushakov IA, Nenajdenko VG. (2008) One-pot synthesis of functionalized indenols from 2-bromoalkenyl trifluoromethyl ketones. Tetrahedron 64:8073–8077.

Short Communication: Postreaction Modifications

Postreaction Modifications of U-4CR-Diels-Alder Products

Alexandra Vorobyeva, Alexey Ilyin, Volodimir Kysil, and Alexandre Ivachtchenko

The postreaction modification of U-4CR products is a well-known method for the synthesis of different heterocyclic systems. The high synthetic productivity of the main reaction (4CR) and the availability of initials make this two-step conversion suitable for the synthesis of both new as well as previously known but not easily accessible heterocycles.

Previously we found an interesting skeleton rearrangement of tandem U-4CR–Diels–Alder products in acid conditions. Further investigation of this reaction sequence demonstrates the influence of the chemical structure of the R1-group upon the type of skeleton rearrangement.

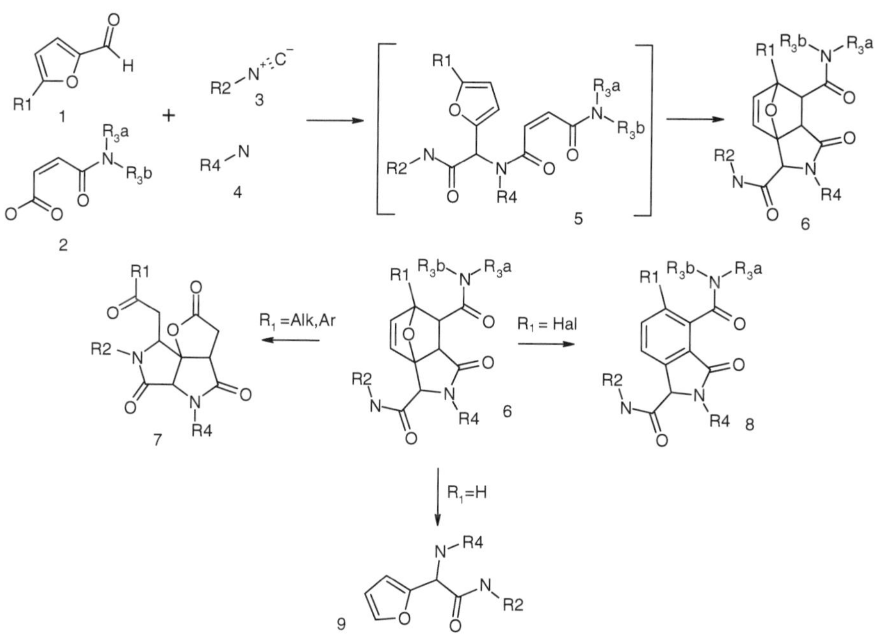

A. Vorobyeva (✉)
Chemical Diversity Research Institute, Khimki, Moscow Reg. 114401, Russia
e-mail: api@chemdiv.com

The R1 alkyl or aryl DA adducts rearrange into the tricyclic saturated system **7**, while in the case of the R1 chlorine- or bromine one the aromatization to **8** occurs. For the products with hydrogen 4s R1 retro-DA reaction and hydrolysis take place.

Some experimental features, proposed mechanisms, and stereochemical details will be discussed.

Session 6
Multi-Component Reactions in Supramolecular Chemistry and Material Science

Multiple Multi-Component Reactions: Rapid Access to Complex Architectures

Ludger A. Wessjohann

Multi-component reactions of building blocks with more than one MCR-reactive group will give rise to oligomeric MCR products. The proper choice of at least two bifunctional building blocks will give either a polymeric or a cyclic product. Apart from polymerization, repetitive or consecutive Ugi reactions have been used to produce linear MCR-heterooligomers with such building blocks.

Through the proper choice of reagents and conditions, however, the (macro) cyclic products can be produced instead. Access to such Multiple Multi-component Macrocycles/Macrocyclizations including Bifunctional Building Blocks (MiBs) will be discussed, with an emphasis on multiple Ugi 4CRs as the most versatile variation (see step 1 in Fig. 1). Polyfunctional building blocks in analogy will allow the construction of three-dimensional polymacrocycles, and dynamic combinatorial libraries can help to select a principal product. If part of the MCR-reactive functional groups is protected, a MiB–Deprotection–MiB (e.g., a MiB–UDC combination; see steps 1–3 in the figure below) can be used to construct almost infinitely

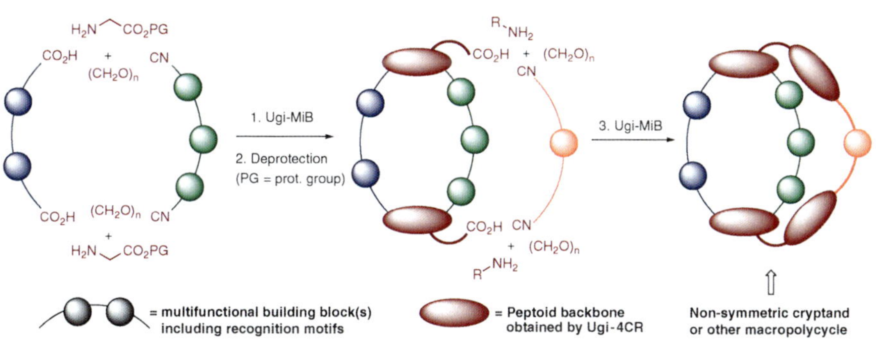

L.A. Wessjohann (✉)
Leibniz Institute of Plant Biochemistry, Halle(Saale), Weinberg 3, D-06120, Germany
e-mail: wessjohann@ipb-halle.de

M.A. Mironov (ed.), *MCR 2009*, Advances in Experimental Medicine and Biology 699,
DOI 10.1007/978-1-4419-7270-5_6,

complex molecular buildings by sequential multiple multi-component reactions. We termed this design and molecular assembly *architectural chemistry*, and some simple as well as advanced examples will be given.

Of course, the use of this strategy is not limited to MiBs as the most complex application. It will be shown that sequential multiple Ugi reactions can used for the construction of more natural linear and cyclic peptide mimics, using a simple ester elongation strategy. This allows access to peptide–peptoid chimeric backbones and, for example, antibiotic peptoids, RGD-peptide mimics, or *N*-methylated cyclopeptides.

References

Wessjohann LA, Rivera DG, Vercillo OE. (2009) Multiple multicomponent macrocyclizations (MiBs): a strategic development toward macrocycle diversity. Chem Rev 109:796–814.

Wessjohann LA, Ruijter E. (2005) Macrocycles rapidly produced by multiple multicomponent reactions including bifunctional building blocks (MiBs). Mol Diversity 9:159.

Rivera DG, Vercillo OE, Wessjohann LA. (2008) Rapid generation of macrocycles with natural-product-like side chains by multiple multicomponent macrocyclizations (MiBs). Org Biomol Chem 6:1787–1795.

Wessjohann LA, Rivera DG, León F. (2007) Freezing imine exchange in dynamic combinatorial libraries with Ugi reactions: versatile access to templated macrocycles. Org Lett 9:4733–4736.

Derivatization of Readily Available meso-Tetraarylporphyrins

Stanisław Ostrowski

Many porphyrin derivatives are of significant importance due to their potential use in many areas of chemistry, medicine, and pharmacology. Usually, the desired precursors for these compounds can be isolated from naturally occurring substances, as the porphyrin system is present in well-known biological materials such as chlorophyll, heme, and vitamin B_{12}. The same effect may be achieved by selective functionalization of the easily available synthetic *meso*-tetraarylporphyrins.

Attempts to prepare compounds possessing a high degree of complexity, by the utilization of the readily available porphyrins, will be presented. Thus, the synthesis, in a fully controlled transformation, of many highly substituted porphyrin systems will be demonstrated, including (a) multi-substituted derivatives in *meso*-aryl rings bearing up to ten functional groups (see path a in the scheme below), (b) double functionalized porphyrins on the same pyrrole unit (path b), (c) exhaustively substituted porphyrins in all β-positions of the so-called Eastern half (path c), (d) simple or more complicated dyads bearing porphyrinyl parts in the structure, among others, furoxans and artificial photosynthetic models (paths d, f).

The intermediates obtained also react with ortho-xylylenes ([4+2]-cycloaddition) and alkyl nitrile oxides ([3+2]-cycloaddition), affording the respective chlorins and isoxazoline-fused chlorins (see examples below; path e).

Some of the above transformations were realized as multi-component reactions.

S. Ostrowski (✉)
Institute of Chemistry, University of Podlasie, ul. 3 Maja 54, 08-110, Siedlce, Poland
e-mail: stan@ap.siedlce.pl

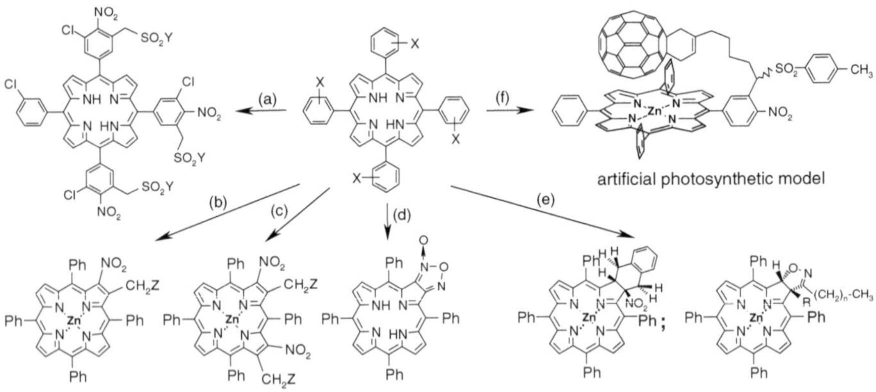

References

Ostrowski S, Mikus A, Łopuszyńska B. (2004) Synthesis of highly substituted *meso*-tetraarylporphyrins. Tetrahedron 60:11951.

Ostrowski S, Mikus A. (2005) Synthesis of porphyrin-acetylene and porphyrin-diene building blocks for new dyads preparation. Heterocycles 65:2339.

Ostrowski S, Wyrębek P, Mikus A. (2006) Synthesis of isoxazoline-fused chlorins by 1,3-dipolar cycloaddition reaction of porphyrins with alkyl nitrile oxides. Heterocycles 68:885.

Ostrowski S, Raczko AM. (2005) Selective double functionalization of *meso*-tetraphenylporphyrin complexes on the same pyrrole unit by tandem electrophilic/nucleophilic aromatic substitution. Helv Chim Acta 88:974.

Synthesis of 7H-Thiazolo[3,2-*a*]pyridines by Three-Component Reaction of Thioamides of Malonic Acid with Acrylonitriles and Dimethyl Acetylenedicarboxylate

Vasiliy Bakulev and Vera Berseneva

We have developed an efficient method for the synthesis of monocyclic, bicyclic, and fused thiazolines based on the reaction of thioamides with dimethyl acetylenedicarboxylate (DMAD). Recently we have found that 3,4-dihydropyridine-2(1*H*)-thiones react with DMAD to form pyridinothiazolines. We have shown that 7*H*-thiazolo[3,2-*a*] pyridines that are interest of medicinal chemistry as potential biologically active compounds can be obtained in high yields by the three-component reaction of malonthioamides with acrylonitriles and DMAD. The structures of the prepared compounds were proved by ^{1}H and ^{13}C NMR spectroscopy and mass-spectrometry. The scope and limitations and the study of the mechanism of the process studied will be reported.

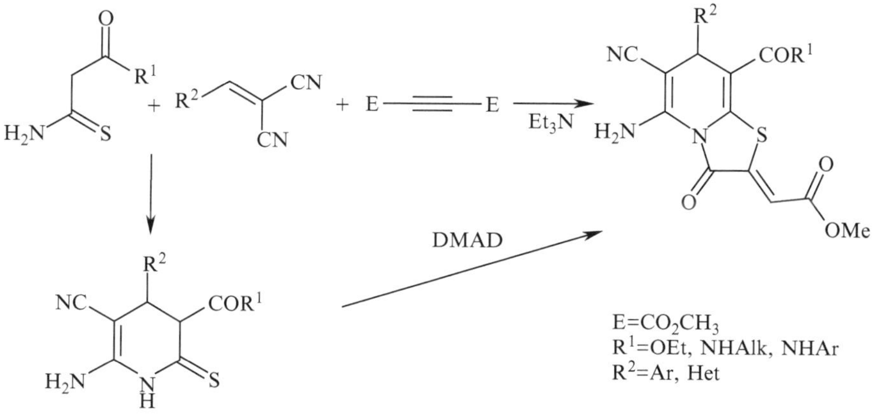

V. Bakulev (✉)
Urals State Technical University, Toslab, Mira 28, Ekaterinburg, 620002, Russia
e-mail: bakulev@mail.ustu.ru

Acknowledgment VAB thanks the Russian Foundation for Basic Research for financial support (grant no. 08-03-00376a).

References

Berseneva VS, Morzherin YY, Dehaen W, Luyten I, Bakulev VA. (2001) Reaction of heterocyclic thioamides with dimethyl acetylenedicarboxylate. Synthesis of novel 2-azolyl-5-methoxycarbonylmethylene thiazolin-4-ones. Tetrahedron 57:2179.

Bakulev VA, Berseneva VS, Belskaia NP, Morzherin YY, Zaitsev A, Dehaen W, Luyten I, Toppet S. (2003) Reactions of 5-mercaptoazoles and pyridine-2-thiones with acetylenic esters. Selectivity of the formation of novel fused thiazin-4-ones and thiazolidin-4-ones. Org Biomol Chem 1(1):134–9.

Multivariance of New Three-Component System Dicarbonyl Compound: Fullerene C_{60}-tris(dialkylamido)phosphite

Andrei V. Bogdanov, Irina P. Romanova, Vladimir F. Mironov, Olga A. Larionova, Gulnara R. Shaikhutdinova, and Oleg. G. Sinyashin

The order of mixing reagents is of significant importance for reactions with more than two reagents. So, a system of a dicarbonyl compound, fullerene C_{60}-tris(dialkylamido)phosphate, proves to be multivariate. Thus, the addition of phosphites (**1, 2**) to fullerene C_{60} leads to the formation of ion-radical salts (**3, 4**); at the same time, the subsequent addition of α-diketones to the reaction mixture does not evoke any reaction. On the other hand, the reaction of phosphite (**2**) with such dicarbonyls as *N*-methylisatin (**5**), acenaphthene- and aceanthrenequinones (**6, 7**) yields only 1,6-diketones (**8–10**), products of the ketocarbenes dimerization. Moreover, the addition of fullerene C_{60} to this reaction solution does not change a reaction result. And the addition of phosphite (**2**) to the mixture of C_{60} with α-diketones (**5–7**) gives only methanofullerenes (**11–13**) with moderate yields. Fullerene acts as a trapping reagent toward the carbene formed in this case.

A.V. Bogdanov (✉)
A.E. Arbuzov Institute of Organic and Physical Chemistry, Russian Academy of Sciences, Arbuzov Street 8, 420088, Kazan, Russia
e-mail: bogdanov@iopc.knc.ru

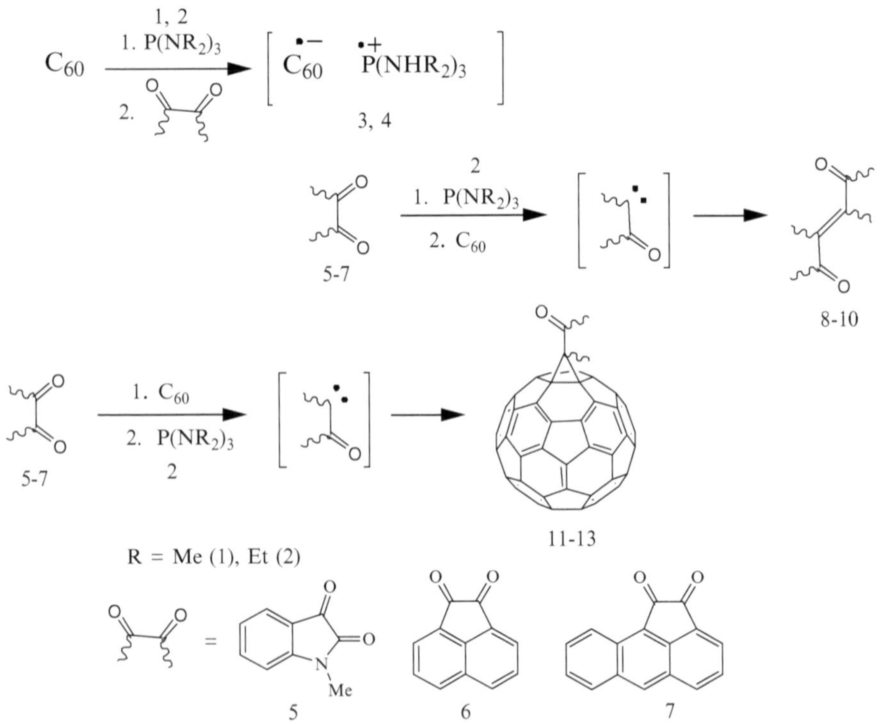

Acknowledgment Financial support by the Russian Foundation for Basic Research and by the Division of Chemistry and Material Sciences of the Russian Academy of Sciences (Program No. 1 и 6) is gratefully acknowledged.

References

Romanova IP, Mironov VF, Larionova OA, Morozov VI, Zverev VV, Sinyashin OG. (2008) Formation of the fullerene radical anion in the reaction of C_{60} with phosphorous triamides. Russ Chem Bull 57:209.

Bogdanov AV, Mironov VF, Buzykin BI, Konovalov AI. (2005) Deoxygenation of acenaphthenequinone with hexaethylphosphorous triamide: an efficient method of synthesis of biacenaphthylidenedione. Russ J Gen Chem 75:825.

Bogdanov AV, Mironov VF, Musin LI, Buzykin BI, Konovalov AI. (2008) Isatin derivatives in the reaction with phosphorous hexaethyltriamide. A new approach to the synthesis of isoindigo derivatives. Russ J Gen Chem 78:1977.

Romanova IP, Bogdanov AV, Mironov VF, Larionova OA, Latypov SK, Balandina AA, Yakhvarov DG, Sinyashin OG. (2009) Fullerene C60 as an effective trap of acenaphthenone carbene generated in the reaction of acenaphthenequinone with hexaethyltriaminophosphine. Mendeleev Commun 19(6):306–308.

One-Pot Synthesis of Nitrogen Heterocycles Under Mannich Reaction Conditions

N. Z. Baibulatova and V. A. Dokichev

A high biological activity of compounds containing piperidine, hexahydropyrimidine, and 3-aza- or 3,7-diazabicyclo[3.3.1]nonane fragments provides a development of novel methods and principles of their production. On the basis of the reaction of 1,3-dinitropropanes, 1,3-dinitrobenzenes, 4-nitropentan-2-one, nitrocarboxylic acid methyl esters, 1,3-acetonedicarboxylic and sulfonediacetic acids' esters and primary amines, we have developed methods of the one-pot synthesis of heterocycles, which can proceed under the conditions of multiple condensation without the production and isolation of intermediates. It was shown that the condensation of the 4-nitrobutane acids' methyl ester with formaldehyde and α,ω,-diaminoalkanes, including lysine methyl ester, occurs to form tricyclic C-18 and C-22 tetraazamacrocycles.

N.Z. Baibulatova (✉)
Institute of Organic Chemistry, Ufa Research Center of the Russian Academy of Sciences, 71 prosp. Oktyabrya, 450054, Ufa, Russia e-mail: chemorg@anrb.ru

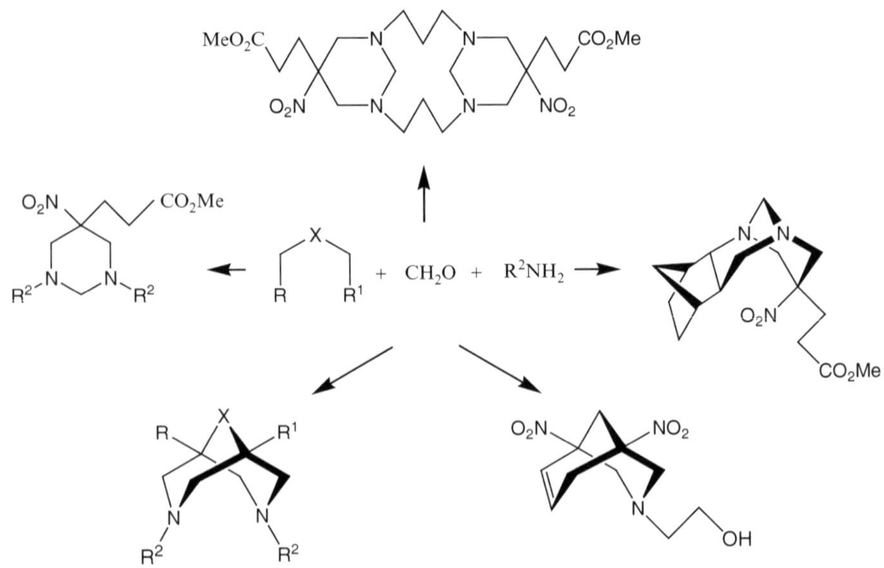

X = CH_2, C=O, SO_2;
R = R^1 = NO_2, C(O)Me, CO_2Me, CO_2Et;
R^2 = Alkyl, Bn, CH_2CH_2OH, $(CH_2)_nNH_2$, $CH(Me)CO_2Me$.

The synthesized compounds have shown antiinflammatory and high antiarrhythmic activity at low toxicity.

Acknowledgment This work was financially supported by the Presidium of the Russian Academy of Sciences (Program for Basic Research "Fundamental Science – Medicine" and "Development of Methodology of Organic Synthesis and Creation of Compound with Valuable Applied Properties").

References

Dokichev VA et al. (1998) Russian patent 2228179.
Dokichev VA et al. (2000) Russian patent 2228334.

Multi-component Synthesis of Pyrrol-Containing Podands Under Heterogeneous Catalysis Conditions

Olga V. Fedorova, Irina G. Ovchinnikova, Gennady L. Rusinov, and Julia A. Titova

For the first time we have shown that the strategy of multi-component condensation is useful for the synthesis of such complicated structures as podands [1]. In this work the high activity of nanosized Al oxides in the multi-component synthesis of substituted pyrrol-containing podands **2**, **4** was demonstrated.

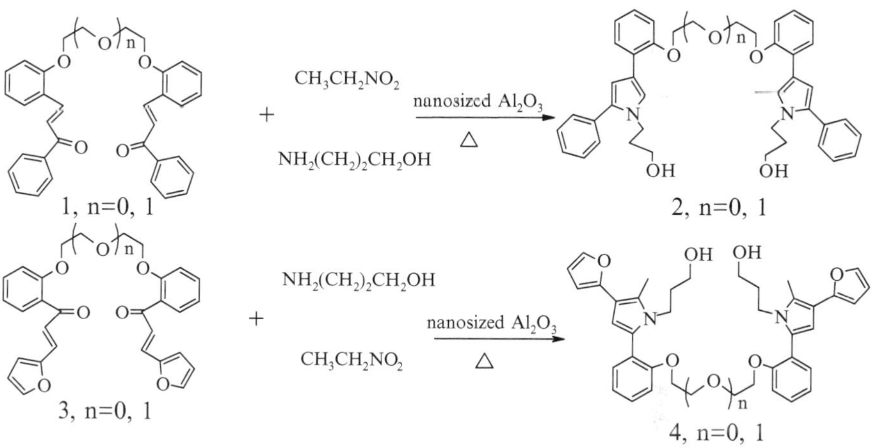

The coordination processes of the reaction participants on the Al nanooxide surface are investigated. We have found that the activation of initial reagents and intermediates results in an increase in both selectivity and the rate of reaction. Thus, the one-pot method of synthesis of compounds **2**, **4** with yields of 55–70% has been developed.

O.V. Fedorova (✉)
I.Ya. Postovsky Institute of Organic Synthesis of Russian Academy of Sciences, Urals Branch, S. Kovalevskoy/Akademicheskay Street, 22/20, 620041, Ekaterinburg, Russia
e-mail: fedorova@ios.uran.ru

Acknowledgment Our work is executed with the financial support of the Russian Foundation for Basic Research (grant nos. 07-03-96111, 07-03-96113); the Presidium of the Russian Academy of Sciences (the projects "Heterogeneous Catalysts with the Use of Nanosized Metals and Their Oxides" and "Design of New Supramolecular Structures Containing Heterocyclic Fragments"), and also the Korea Atomic Energy Research Institute (contract 01/06).

Reference

Zhidovinova MS, Fedorova OV, Rusinov GL, Ovchinnikova IG. (2003) Multicomponent sonochemical synthesis of podands. Mol Diversity 6:323–326.

Analysis on Hydrogen and Methane Combustion Reactions and Particle Transport in Premixed Flame Reactor

Anna Nasonova, Dong-Joo Kim, Kyo-Seon Kim, Young Min Jang, and Sung Joon Kim

In this study, we use the commercial CFD-code FLUENT to analyze the hydrogen and methane combustion in the premixed flame reactor numerically. The three-coaxial burner is used: H_2 or CH_4 mixed with air flows through the central tube of the burner; H_2 through the middle annulus; and air through the outer annulus. The computational domain is confined by a coaxial pyrex tube (diameter = 5.2 cm, height = 30 cm) surrounding the flame. The rate of hydrogen-to-air/methane-to-air mixture combustion is calculated as the Arrhenius rate and the mixing rate.

We calculated the profiles of fluid flow and temperature for various process conditions. To calculate particle trajectories inside the premixed reactor, we utilized data of temperature and velocity profiles from FLUENT. We found that particles starting at different initial positions move inside the reactor with different particle trajectories. Inside the flame reactor, the gradient of temperature causes the particles to move in the direction of decreasing temperature, in a phenomenon known as thermophoresis. The movement of particles along their trajectories in the premixed flame reactor is strongly affected by the gas velocity and thermophoresis. The particle trajectory investigation is important to predict particle properties because particles following different trajectories might experience different gas temperatures in the flame reactor.

References

Yu MZ, Lin JZ, Chan T. (2008) Numerical simulation of nanoparticle synthesis in diffusion flame reactor. Powder Technol 181(1):9.

Mendes MAA, Pereira JMC, Pereira JCF. (2008) A numerical study of the stability of one-dimensional laminar premixed flames in inert porous media. Combust Flame 153:525.

Sunsap P, Kim D-J, Charinpanitkul T, Kim W-S, Kim K-S. (2008) Computational analysis on fluid dynamics during the flame synthesis of TiO_2 nanoparticles. J Korean Phys Soc 52(4):1298.

K.-S. Kim(✉)
Department of Chemical Engineering, Kangwon National University, Hyoja-2-dong, Chuncheon, Korea, 200-701
e-mail: kkyoseon@kangwon.ac.kr

Plasma Reactions of NO and SO_2 Removal in Dielectric Barrier Discharge: Photocatalysts Hybrid System

Anna Nasonova, Dong-Joo Kim, and Kyo-Seon Kim

We analyzed the NO and SO_2 removal in the dielectric barrier discharge–photocatalysts hybrid system. The dielectric barrier discharge for NO and SO_2 removal was generated by dielectric barrier discharge with glass beads as a dielectric material. Glass beads were coated with TiO_2 thin film by using a rotating cylindrical plasma chemical vapor deposition (PCVD) reactor. The PCVD reactor is the quartz cylinder with an inner diameter and length of 55 and 280 mm, respectively. The inductively coupled plasmas are generated by applying the electric rf power to a water-cooled spiral-shaped coil electrode that is located outside the cylindrical reactor. The precursor for the TiO_2 thin-film formation is titanium tetraisopropoxide (TTIP). The PCVD process allows production of the dense and uniform TiO_2 thin films on the glass beads' surfaces.

In the low-temperature plasma process, the energetic electrons are generated by the high-voltage electric power supply. The active radicals for NO and SO_2 removal (OH, HO_2, O, N, etc.) are produced by the electron collision reactions with gas molecules. Those radicals oxidize SO_2 and NO to form acid molecules such as H_2SO_4 and HNO_3, respectively. The acid molecules are chemically neutralized by the reactions with ammonia (NH_3) and converted into ammonium sulfate [$(NH_4)_2SO_4$] and ammonium nitrate (NH_4NO_3), respectively. The NO and SO_2 removal efficiencies increase with increases in applied peak voltage. As the residence time increases or the initial NO and SO_2 concentrations decrease, the NO and SO_2 removal efficiencies increase. The NO and SO_2 removal efficiencies obtained in a nonthermal plasma reactor packed with TiO_2-coated glass beads were higher than those obtained when uncoated glass beads were used.

K.-S. Kim (✉)
Department of Chemical Engineering, Kangwon National University,
Hyoja-2-dong, Chuncheon, Korea, 200-701
e-mail: kkyoseon@kangwon.ac.kr

References

Filimonova EA, Amirov RH, Kim HT, Park IH. (2000) Comparative modeling of NO_x and SO_2 removal from pollutant gases using pulsed-corona and silent discharges. J Phys D Appl Phys 33:1716.

Penetrante BM, Brusasco RM, Merritt BT, Vogtlin GE. (1999) Environmental applications of low-temperature plasmas. Pure Appl Chem 71(10):1829.

Dors M, Mizeraczyk J. (2004) NO_x removal from a flue gas in a corona discharge-catalyst hybrid system. Catal Today 89:127.

Chemical Modification of Chitin and Chitosan by Formaldehyde and H_2S

E. B. Rakhimova, A. B. Minnebaev, and V. R. Akhmetova

Polysaccharide materials based on chitin, chitosan, and its derivatives may be successfully used for ion-exchange membranes in ultrafiltration and dialysis. Moreover, the complex-forming ability of these biopolymers can be used for selective extraction of several metals from seawater.

It is known that the modification of chitosan **1** by formaldehyde and acetaldehyde occurs with the formation of Schiff bases (scheme). However, multi-component reactions of chitin **2** and chitosan **1** with aliphatic aldehydes and hydrogen sulfide were not studied.

We first carried out the chemical modification of biopolymers **1** and **2** using thiomethylating reagent "CH_2O–H_2S." This multi-component reaction proceeds in two directions: (1) intramolecular heterocyclization at 20°C and (2) intermolecular cross-linkage at 60°C. As a result, "hybrid" polysaccharides containing in its chain O-, N-, and S-macroheterocycles **1a** (S 9.67%), **2a**, and polysaccharides cross-linked by methylensulfide chains **1b** (S 15.60%) were obtained in 70–75% yield. The structures of **1a**, **1b**, and **2a** were established by IR-, ^{1}H, ^{13}C NMR-spectra (COSY, HSQC, HMBC) and by elemental analysis.

Scheme

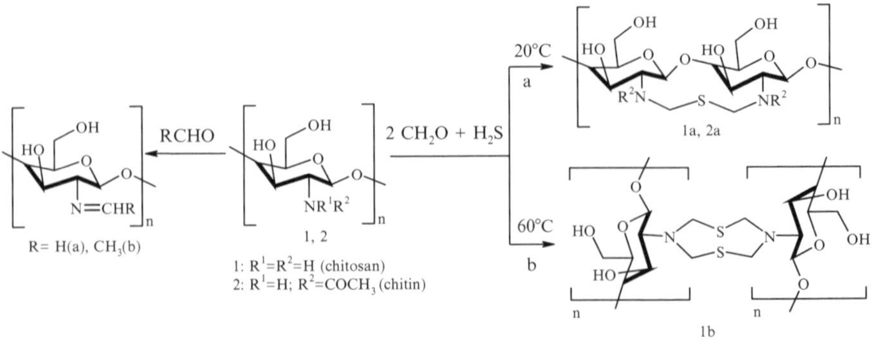

V.R. Akhmetova (✉)
Institute of Petrochemistry and Catalysis of the Russian Academy of Sciences, 450075, Ufa, Russia
e-mail: ink@anrb.ru

Sulfur-containing derivatives of chitin and chitosan **1a**, **1b**, and **2a** may be of interest as highly efficient extragents and sorbents of noble and rare metals, as well as of selective complex ones.

Reference

Kurita K, Mori S, Nishiyama Y, Harata M. (2002) N-Alkylation of chitin and some characteristics of the novel derivatives. Polym Bull 48:159–166.

Structure Diversification and Combinatorial Synthesis of Cyclodecapeptides Based on Natural Laterocidin

Chuanguang Qin, Ruijie Zhang, Qiuyu Wang, Jin Ren, and Linqi Tian

Cyclopeptides represent a large and interesting class of bioactive molecules displaying a broad spectrum of significant biological activities. The synthesis of head-to-tail cyclopeptides has attracted considerable interest since the antibiotic gramicidin S was found to be a cyclodecapeptide. Many antibiotics and toxins are also known to be cyclopeptides. Due to the growing interest that cyclic peptides have generated, intense efforts have been invested in developing efficient methods for peptide cyclization.

Laterocidin is a novel cyclodecapeptide antibiotic isolated from laboratory cultures of a bacterial strain (*Brevibacillus laterosporus* VKPM B-8287). It exhibits distinctive potent antibiotic activities among the family of cyclopeptide antibiotics that includes gramicidin S, tyrocidines, streptocindins, and loloatins. To facilitate the structure–activity relationship study, novel and facile synthetic methods are highly desirable although these natural products have been synthesized by traditional methods. Based on structural characteristics of natural laterocidin (see Fig. 1), here we report a novel method for total solid-phase synthesis of laterocidin and its analogs by using the Dmab group as a temporary α-COOH protecting group during solid-phase synthesis with Fmoc chemistry. This synthetic method for the natural products will not only expedite the elucidation of the structure–activity relationships, but will also significantly facilitate the optimization of their therapeutic index for containment of microbial resistance.

C. Qin(✉)
Faculty of Life Science, Northwestern Polytechnical University, Xi'an, 710072, China
e-mail: qinchg@nwpu.edu.cn

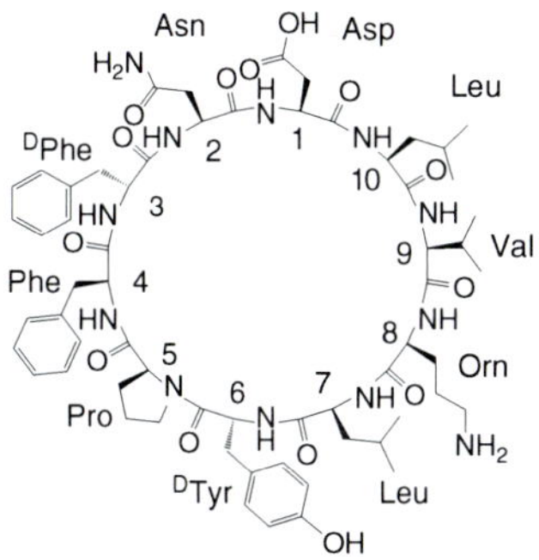

1. Cyclo[-LVOL(^{D}Y)PF(^{D}F)ND-]
2. Cyclo[-LVOL(^{D}Y)PF(^{D}F)NA-]
3. Cyclo[-LVOL(^{D}Y)PF(^{D}F)AD-]
4. Cyclo[-LVOL(^{D}Y)PF(^{D}A)ND-]
5. Cyclo[-LVOL(^{D}Y)PA(^{D}F)ND-]
6. Cyclo[-LVOL(^{D}Y)AF(^{D}F)ND-]
7. Cyclo[-LVOL(^{D}A)PF(^{D}F)ND-]
8. Cyclo[-LVOA(^{D}Y)PF(^{D}F)ND-]
9. Cyclo[-LVAL(^{D}Y)PF(^{D}F)ND-]
10. Cyclo[-LAOL(^{D}Y)PF(^{D}F)ND-]
11. Cyclo[-AVOL(^{D}Y)PF(^{D}F)ND-]
12. Cyclo[-LVOL(^{D}Y)PF(^{D}F)NN-]
13. Cyclo[-LVOL(^{D}F)PF(^{D}F)NN-]
14. Cyclo[-LVOL(^{D}F)PY(^{D}F)NN-]
15. Cyclo[-LVOL(^{D}F)PYFNN-]
16. Cyclo[-LVOLFPY(^{D}F)NN-]
17. Cyclo[-LVOLFPYFNN-]

Fig. 1 Structure of natural laterocidin and sequences of synthetic analogs

Acknowledgment This work was financially sponsored by the National Natural Science Foundation of China (NSFC 20672086) and the Scientific Research Foundation for the Returned Overseas Chinese Scholars, State Personnel Ministry of China (2007).

Reference

Qin CG et al. (2003) Dissociation of antibacterial and hemolytic activities of an amphipathic peptide antibiotic. J Med Chem 46:4830–4833.

Modification of Small Chitosan Particles via Multi-Component Reactions

Maxim A. Mironov

During the last 5 years, our research group has concentrated its efforts on multi-component reactions in water and aqueous solutions. Several interesting effects, including the acceleration of isocyanide-based multi-component reactions in pure water, have been found. In the course of this project, several novel water-soluble isocyanides have been synthesized. In this study, we tried to apply the obtained results for the development of encapsulation technology.

Chitosan is natural polysaccharide that is issued from chitin via partial deacetylation. In the field of water-soluble polymers, chitosan has earned growing interest due to these unique properties, including biodegradability and biocompatibility. Therefore, this polymer is a very promising candidate for the development of controlled-release systems such as micro- and nanocapsules. As a rule, chitosan is dissolved via the addition of hydrochloric or acetic acids, so the true solution can be used in many applications, for example, in the formation of polyionic complexes on the surface of emulsion droplets. Another way to obtain aqueous solutions of chitosan is from colloidal systems containing small charged particles with a diameter of 150–300 nm. However, the thermodynamically unstable nature of colloidal systems is a critical issue in all applications. For example, colloidal solutions of chitosan are stable only in a certain pH range (6.7–7.1). The modification of small particles results in stabilization of the systems, which is a way to control the agglomeration and formation of thin films on the surface of emulsion droplets.

In general, our approach includes three stages: (1) obtaining chitosan particles with a diameter of 0.3–0.5 μm via simple coacervation of chitosan hydrochloride; (2) modification and partial hydrophobization of the particle surface via multi-component reaction: chitosan, levulinic acid, and isocyanide; (3) formation of microcapsules. As a result, capsules having various porosities have been obtained. The possibilities for this material in the design of controlled-release systems will be presented.

M.A. Mironov (✉)
Department of Chemistry, TOSLab, Urals State Technical University,
Mira Street 19, 620002, Ekaterinburg, Russia
e-mail: mironov@mail.ustu.ru

References

Mironov MA, Ivantsova MN, Mokrushin VS. (2006) A novel isocyanide-based multicomponent reaction: an easy access to substituted propionamides and succinimides. Synlett 4:615–618.

Mironov MA, Ivantsova MN, Tokareva MI, Mokrushin VS. (2005) Acceleration of the Passerini reaction in the presence of nucleophilic additives. Tetrahedron Lett 46:3957–3959.

Mironov MA, Ivantsova MN, Mokrushin VS. (2003) Ugi reaction in aqueous solutions: a simple protocol for libraries production. Mol Divers 6:193–198.

Short Communications: Synthesis Reactions

Solid-State Synthesis of Chitosan-Polyester Systems for Biomedical Applications

Tatiana S. Demina, Anna M. Tcoy, Tatiana A. Akopova, Elena A. Markvicheva, Alexander O. Chernyshenko, and Alexander N. Zelenetski

Chitosan is a natural cationic linear polymer that is currently receiving a great deal of interest for medical and pharmaceutical applications. Chitosan is biodegradable and bioadhesive and has bacteriostatic effects. The use of chitosan as a biomaterial that can support chondrogenesis has drawn considerable attention in applications for the repair of articular cartilage. Chemical derivatization of chitosan, including grafting with synthetic polymers, provides a powerful means to promote new biological activities and to modify its mechanical properties. Poly(L-lactic acid) (PLLA) is one of the synthetic polymers often used to design materials for tissue engineering.

The novel approach to new polysaccharide-based materials in a simple one-step procedure under conditions of shear deformation, solid-state reactive blending (SRB), was employed to obtain chitosan-PLLA systems. The entire modification process proceeds in the solid state of the components and does not require catalysts or any solvents as reaction medium. This is desirable for biomedical applications and provides numerous possibilities to circumvent many processing obstacles typical for the preparation of natural-polymer-based materials. The compositions at a three-to-one molar ratio of PLLA to chitosan were obtained by SRB in an extruder at various temperatures (below the PLLA's melting point). As follows from the DSC data, the maximal structure transformations resulting in partial amorphization of the blends happen in a rather low temperature range, whereas a temperature rise leads to recrystallization as well as to chemical reactions between the functional groups of these polymers. Fractional analysis of the blends and the study of separated fractions by the FTIR spectroscopic method revealed that the salt linkages between the end carboxyl groups of PLLA and amino-groups of chitosan are formed for the blends obtained at 100°C, whereas at higher blending temperatures a fraction of the graft copolymers is formed. The grafting proceeds through the amidation of the end carboxyl groups of polyester by amino-groups of chitosan as well as due to reactions of alminolysis and alcoholysis of the ester bonds by the

T.S. Demina (✉)
Enikolopov Institute of Synthetic Polymer Materials, Russian Academy of Sciences, 70, Profsousnay Street, 117393, Moscow, Russia
e-mail: detans@gmail.com

functional groups of chitosan. In organic solvents these blends are distributed on a colloidal level, forming dispersions of micelle-like aggregates of a size of 750–800 nm.

The application of obtained materials to biomedical problems such as cell cultivation in vitro in preparation to implant them later to promote tissue regeneration was investigated. The system is based on 40%-wt of chitosan processed into microparticles by the emulsion/evaporation technique for their next usage as microcarriers. The attachment of fibroblasts (L929) on the surface of microparticles was observed and the number of cells was increased during incubation.

Amberlyst A-15: An Efficient Reusable Catalyst for the Synthesis of 2,4,5-tri-Arylimidazole and 1,2,4,5-tetra-Arylimidazole Derivatives Under MW Irradiation

Shivaji S. Pandit

Heterocyclic compounds with imidazole moiety have many pharmaceutical activities and play important roles in biochemical processes. Imidazole ring systems are frequently found in numerous naturally occurring and synthetic molecules, such as those with herbicidal, antiviral, antifungal, and antihelmitic properties. The imidazole ring system is of particular interest as it is a component of histidine that produces histamine in metabolic processes. Triarylimidazoles are found in photography as photosensitive compounds. Despite their importance from pharmacological, industrial, and synthetic points of view, comparatively few methods have been reported. These methods include the condensation of 1,2-diketones, aldehydes, and ammonium acetate in the presence of H3PO4, H_2SO_4, AcOH, SSA, and *p*-TSA. Recently, one-pot condensation of an aldehyde and ammonium acetate with α-hydroxy ketones, α-keto-oximes, or 1,2-diketones has been performed using solid support under molecular weight (MW). These methods have a number of drawbacks, including poor yields, excessive use of reagents, high temperature (180–200°C), high cost of conventional room-temperature ionic liquids, use of toxic organic solvent, and use of excess of ammonium acetate.

In connection to our trend toward green chemistry and the biological importance of substituted imidazole derivatives, we have developed an ecologically safe strategy for the synthesis of imidazole derivatives in the presence of Amberlyst A-15 as reusable catalysts under MW irradiation.

S.S. Pandit (✉)
Department of Chemistry, Arts, Science and Commerce Kolhar, Tal. Rahata District, Ahmednagar, 413710, Maharashtra, India
e-mail: akankshapandit2002@yahoo.com

Conference Program

All lectures were held in the Conference Hall of the Main Building of the Urals State Technical University (Mira Str., 19).

Monday, 25 May

Session 1 **Catalysis and MCR** *Chairman*: **Romano Orru**

08.45–09.10	*Conference opening*	
09.10–10.00	**Prof. Masakatsu Shibasaki** *The University of Tokyo, Japan*	Recent Progress in Asymmetric Two-Center Catalysis
10.00–10.35	**Prof. Miguel Yus** *Universidad de Alicante, Spain*	Efficiency in Chemistry: From Hydrogen Autotransfer to Multi-component Catalysts
10.35–10.55	*Coffee break*	
10.55–11.30	**Prof. Michael Organ** *York University, Toronto, Ontario, Canada*	Multi-component Reactions (MCR) by Microwave-Assisted, Continuous Flow Organic Synthesis (MACOS)
11.30–12.05	**Prof. Erik Van der Eycken** *Catholic University of Leuven, Belgium*	A Potpourri of Recent Microwave-Assisted 2(1*H*)-Pyrazinone Chemistry
12.05–12.25	**Eugen Merkul** *Heinrich-Heine-Universität, Duesseldorf, Germany*	Efficient Construction of (Hetero) Aryl Substituted (Aza)Indoles by Pd-Catalyzed Multi-Component Reactions
12.25–12.45	**Sergey V. Ryabukhin** *Enamine Ltd., Kiev, Ukraine*	Chlorotrimethylsilane Iis an Efficient Promoter for the Biginelli Reaction
12.45–12.55	**Millab** *presentation*	
12.55–14.00	*Lunch break*	

Session 2 **MCR in Heterocyclic Chemistry** *Chairman*: **Jieping Zhu**

14.00–14.50	**Prof. Thomas Müller** *Heinrich-Heine-Universität, Düsseldorf, Germany*	Multi-component Synthesis of Heterocycles by Palladium-Catalyzed Generation of Alkynones, Alkenones, and Allenes
14.50–15.25	**Prof. Romano Orru** *Vrije University, Amsterdam, The Netherlands*	Multi-component Reactions as Useful Platforms to Explore the Chemical Space

15.25–15.45	*Coffee break*	
15.45–16.20	**Dr. Alexey Ilyin** *Chemical Diversity Research Institute, Moscow, Russia*	Polyfunctional U-MCR Reagents: Convenient Pathway to Novel Heterocyclic Systems
16.20–16.55	**Dr. Valentine Chebanov** *Institute for Single Crystals, Kharkov, Ukraine*	Multi-component Heterocyclizations: Control of Chemo- and Regioselectivity
16.55–17.15	**Prof. Vnira R. Akhmetova** *Institute of Petrochemistry and Catalysis of Russian Academy of Sciences, Ufa, Russia*	Multi-component Reactions of H_2S with Aldehydes and Amines as Efficient Route to Heterocycles and Thioaza-Crown Compounds
17.15–17.35	**Dr. Leonid G. Voskressensky** *Russian Peoples' Friendship University, Moscow, Russia*	Alkynes-Induced Tandem Transformations of (Hetero) Annulated 5-, 6- or 7-Membered *N*-Heterocycles and a New MCR on This Motif
17.35–17.55	**Prof. Andrey N. Maslivets** *Perm State University, Perm, Russia*	Cascade Recyclizations of 1*H*-Pyrrole-2,3-diones: Method of Unusual Heterocyclic Systems Construction
17.55–18.40	*Poster session 1 (Posters PS1–PS26)*	

Tuesday, 26 May

Session 3 **MCR in Drug Discovery** *Chairman*: **Luca Banfi**

09.00–09.50	**Prof. Christopher Hulme** *Pharmacy-Skaggs Center, University of Arizona, USA*	MCRs: Bench to Bedside with "Iterative Efficiency"
09.50–10.25	**Dr. Irini Akritopoulou-Zanze** *Abbott Laboratories, Medicinal Chemistry Technologies, USA*	Multi-component Reactions and Drug Discovery
10.25–10.45	*Coffee break*	
10.45–11.20	**Prof. Alexander Dömling** *Pharmaceutical Sciences, University of Pittsburgh, USA*	A Fast and Efficient Approach to Small Molecular Weight Protein–Protein Interaction Antagonists Using MCR Chemistry
11.20–11.55	**Prof. Wei-Min Dai** *The Hong Kong University of Science and Technology, Hong Kong SAR, China*	Building Molecular Shape Diversity via Post-Ugi Annulations
11.55–12.15	**Dr. Volodimir Kysil** *ChemDiv., San Diego, CA, USA*	Isocyanide-Based Multi-Component Reaction of Diamines: An Efficient Strategy for Assembling of Unique Heterocyclic Scaffolds
12.15–12.35	**Dr. Cedric Kalinski** *Priaxon AG, Munich, Germany*	Priaxon AG, a New Dimension in Drug Discovery
12.35–12.55	**Prof. Chuanguang Qin** *Northwestern Polytechnic University, Xi'an, China*	Molecular Diversification and Solid-Phase Synthesis of *N*-Acylpeptides for Antibiotic Screening
12.55–14.00	*Lunch break*	

Session 4 **Novel Reagents for MCR** *Chairman*: **Vasiliy Bakulev**

14.00–14.50	**Prof. Wang Yanguang** *Zhejiang University, Hangzhou, China*	Ketenimine Intermediate-Based Multi-component Reactions
14.50–15.25	**Dr. Andrea Basso** *Università di Genova, Genova, Italy*	Pluripotent Substrates in Diversity-Oriented Synthesis
15.25–15.45	*Coffee break*	
15.45–16.20	**Prof. Valentine Nenajdenko** *Moscow State University, Moscow, Russia*	New Polyfunctional Reagents for Isocyanide-Based Multi-component Reactions
16.20–16.40	**Prof. Renata Riva** *Università di Genova, Genova, Italy*	Highly Functionalized Tricyclic *N*-Heterocycles by Coupling an Ugi MCR with a Tandem Pd(0)-Promoted S_N2'–Heck Reaction
16.40–17.00	**Dr. Mikhail Krasavin** *Chemical Diversity Research Institute, Khimki, Russia*	Products of Ugi Reaction Among Acid Hydrazides, Aldehydes, and Isocyanides with Added Diversity
17.00–17.20	**Prof. Song Cao** *East China University of Science and Technology, Shanghai, China*	New Strategy for the Synthesis of Fluorinated Compounds: Multi-component Reactions Including Fluoro-Containing Building Blocks (MCRs-FBB)
17.20–17.40	**Dr. Igor B. Rozentsveig** *A.E. Favorsky Irkutsk Institute of Chemistry, Siberian Branch of Russian Academy of Sciences, Irkutsk, Russia*	Cascade Transformations of *N*-(1-aryl-2-Polychloroethyl) Sulfonamides
17.40–18.00	**Prof. Lu Ping** *Zhejiang University, Hangzhou, China*	A Multi-component Reaction Leading to a Phenalenone Derivative
18.00–18.45	*Poster session 2 (Posters PS27–PS52)*	

Wednesday, 27 May

Session 5 **Design of MCR** *Chairman*: **Maxim Mironov**

09.00–09.50	**Prof. Jieping Zhu** *Institut de Chimie des Substances Naturelles, France*	Isocyanide-Based Multi-component Reactions: New Variations
09.50–10.25	**Prof. Laurent El Kaïm** *Unité Chimie et Procédés, ENSTA, Paris, France*	From Ugi–Smiles Couplings to Nef Reaction of Isocyanides
10.25–10.45	*Coffee break*	
10.45–11.20	**Dr. Gian Cesare Tron** *DISCAFF, Novara, Italy*	The Importance of Being a Secondary Amine. From the *N*-split Ugi Reaction to the Reshuffling of the Ugi and Passerini Products
11.20–11.55	**Prof. Rodolfo Lavilla** *Institute for Research in Biomedicine, Barcelona, Spain*	Rebuilding a Forgotten MCR: Studies on the Interaction of Isocyanides with Alkylboranes and Aldehydes

11.55–12.15	**Prof. Luca Banfi** *Università di Genova, Genova, Italy*	Diversity-Oriented Synthesis of Various Benzo-Fused Oxaza Heterocycles by Coupling the Ugi Reaction with Nucleophilic Substitutions
12.15–12.35	**Nicola Kielland** *Institute for Research in Biomedicine, Barcelona, Spain*	New Multi-component Reactions Based on Electrophilic Activation of Azines
12.35–12.55	**Prof. Alevtina S. Medvedeva** *A.E. Favorsky Irkutsk Institute of Chemistry, Siberian Branch of Russian Academy of Sciences, Irkutsk, Russia*	Multi-component Synthesis of New Heterocyclic Compounds from Propynals
12.55–13.15	**Prof. Yurii V. Shklyaev** *Institute of Technical Chemistry of Urals Branch of Russian Academy of Sciences, Perm, Russia*	Synthesis of Partially Hydrogenated Heterocycles: 3,4-Dihydroisoqunolines, Spiropyrrolines, Polyhydroindols
13.15–14.00	**Reaxys workshop (Elsevier)** **Dr. Eva Seip** *Elsevier MDL GmbH, Germany*	An Introduction to Reaxys: The New Workflow Solution for Research Chemists
15.00–21.00	*Excursion and conference dinner*	

Thursday, 28 May

Session 6 **MCR in Supramolecular Chemistry and Material Science**
Chairman: **Valentine Nenajdenko**

09.00–09.50	**Prof. Ludger Wessjohann** *Leibniz Institute of Plant Biochemistry, Germany*	Multiple Multi-component Reactions: Rapid Access to Complex Architectures
09.50–10.25	**Prof. Stanisław Ostrowski** *University of Podlasie, Siedlce, Poland*	Derivatization of Readily Available *meso*-Tetraarylporphyrins
10.25–10.45	**Dr. Andrei V. Bogdanov** *A.E. Arbuzov Institute of Organic and Physical Chemistry, Kazan, Russia*	Multivariance of New Three-Component System Dicarbonyl Compound – Fullerene $_{60}$ – Tris(dialkylamido)phosphate
10.45–11.05	**Tatiana S. Demina** *Enikolopov Institute of Synthetic Polymer Materials, Russian Academy of Sciences, Moscow, Russia*	Solid-State Synthesis of Chitosan-Polyester Systems for Biomedical Applications
11.05–11.25	**Prof. Vasily Bakulev** *Urals State Technical University, Ekaterinburg, Russia*	Synthesis of 7*H*-Thiazolo[3,2-*a*] pyridines by Three-Component Reaction of Thioamides of Malonic Acid with Acrylonitriles and Dimethyl Acetylenedicarboxylate
11.25–11.45	**Dr. Shivaji S. Pandit** *Arts, Science and Commerce College, Kolhar, India*	Amberlyst A-15: An Efficient Reusable Catalyst for the Synthesis of 2, 4, 5-*tri*-Arylimidazole and 1,2,4,5-*tetra*-Arylimidazole derivatives under MW Irradiation
11.45–12.00	*Conference closure*	

List of Participants

Prof. Vnira Rakhimovna Akhmetova
Institute of Petrochemistry and Catalysis of Russian Academy of Sciences, Ufa, Russia
E-mail: vnirara@mail.ru; ink@anrb.ru

Dr. Irini Akritopoulou-Zanze
Abbott Laboratories, Medicinal Chemistry Technologies, Abbott Park, IL, USA
E-mail: Irini.Zanze@abbott.com

Prof. Alexander Vladimirovich Anisimov
Moscow State University, Moscow, Russia
E-mail: sulfur45@mail.ru

Jacob Ansumana
Siky System Production, Freetown, Sierra Leone
E-mail: saheedmohamed@yahoo.com

Prof. Vasiliy Alekseevich Bakulev
Urals State Technical University, Russia
E-mail: bakulev@r66.ru

Prof. Luca Banfi
Università di Genova, Dipartimento di Chimica e Chimica Industriale, Genova, Italy
E-mail: banfi@chimica.unige.it

Sahareh Bashiribod, M.S.
Shahid Behashty University, Tehran, Iran
E-mail: sahareh.b@gmail.com

Dr. Andrea Basso
Dipartimento di Chimica e Chimica Industriale, Università degli Studi di Genova, Genova, Italy
E-mail: andreab@chimica.unige.it

Nailya Zinurovna Baybulatova
Institute of Organic Chemistry, Ufa Research Center of the Russian Academy of Sciences, Ufa, Russia
E-mail: dokichev@anrb.ru

Soheila Behzadi, M.S.
Department of Chemistry, Shahid Bahonar University of Kerman, Kerman, Iran
E-mail: sohela_behzadi@yahoo.com

Dr. Nikolay Arkadievich Belayev
Urals State Technical University, Ekaterinburg, Russia
E-mail: belayev@mail.ustu.ru

Francesca Bevacqua
Università degli Studi di Genova/Pharmacy S. Tommaso, Genova, Italy
E-mail: andreab@chimica.unige.it

Valentina Bevilacqua
Università del Piemonte Orientale, Novara, Italy
E-mail: bevilacquav@gmail.com

Dr. Andrey Vladimirovich Bogdanov
A.E. Arbuzov Institute of Organic and Physical Chemistry Kazan Scientific Center, Russian Academy of Sciences, Kazan, Russia
E-mail: bogdanov@iopc.knc.ru

Dr. Andriy Boholyubsky
Enamine Ltd., Kiev, Ukraine
E-mail: bogolubsky@mail.enamine.net

Olga Bulatova
IBC Corporate Travel, Ekaterinburg, Russia
E-mail: pr@ibc.ru

Ekaterina Bushkova
Chemical Diversity Research Institute, Khimki, Russia
E-mail: bushkova@iihr.ru

Tatiana Sergeevna Bushneva
Millab, Moscow, Russia
E-mail: analytik@millab.ru

Prof. Valery Nikolaevich Charushin
I.Ya. Postovsky Institute of Organic Synthesis, Urals Branch of Russian Academy of Sciences, Ekaterinburg, Russia
E-mail: charushin@ios.uran.ru

Dr. Valentyn Chebanov
State Scientific Institution "Institute for Single Crystals" NAS of Ukraine, Kharkiv, Ukraine
E-mail: chebanov@isc.kharkov.com

Dr. Chea Louis Chefuh
ENSP Yaounde, Yaounde, Cameroon
E-mail: louisbourg11@yahoo.fr

Maria Andreevna Cherepanova
Urals State Technical University-UPI, Ekaterinburg, Russia
E-mail: mcr2009@mail.ustu.ru

Prof. Oleg Nikolaevich Chupakhin
I.Ya. Postovsky Institute of Organic Synthesis, Urals Branch of Russian Academy of Sciences, Ekaterinburg, Russia
E-mail: chupakhin@ios.uran.ru

Prof. Minoo Dabiri
Department of Chemistry, Faculty of Science, Shahid Beheshti University, Tehran, Iran
E-mail: m-dabir@sbu.ac.ir

Prof. Wei-Min Dai
Department of Chemistry, The Hong Kong University of Science and Technology, Clear Water Bay, Kowloon, Hong Kong SAR, China
E-mail: chdai@zju.edu.cn; chdai@ust.hk

Dr. Gevorg G. Danagulyan
Russian-Armenian (Slavonic) State University, Institute of Organic Chemistry, NAS RA, Yerevan, Republic of Armenia
E-mail: gdanag@email.com

Prof. Alexander Dömling
Departments of Pharmacy and Chemistry, Drug Discovery Institute, University of Pittsburgh, Pittsburgh, PA, USA
E-mail: asd30@pitt.edu

Dr. Fabio De Moliner
Università di Genova, Dipartimento di Chimica e Chimica Industrial/ MerckSerono, Genova, Italy
E-mail: fabio.demoliner@libero.it

Prof. Maria del Carmen
University of Alicante, Alicante, Spain
E-mail: yus@ua.es

Tatiana Sergeevna Demina
Enikolopov Institute of Synthetic Polymer Materials, Russian Academy of Sciences, Moscow, Russia
E-mail: detans@gmail.com

Dr. Mohammadhossein Farjam
Islamic Azad University, Firoozabad, Iran
E-mail: hosein_farjam@yahoo.com

Dr. Olga V. Fedorova
I.Ya. Postovsky Institute of Organic Synthesis of Russian Academy of Sciences, Urals Branch, Ekaterinburg, Russia
E-mail: fedorova@ios.uran.ru

Prof. Alexander S. Fisyuk
F.M. Dostoevsky Omsk State University, Omsk, Russia
E-mail: fis@orgchem.univer.omsk.su

Anna K. Garkushenko
F.M. Dostoevsky Omsk State University, Omsk, Russia
E-mail: sagitullina@orgchem.univer.omsk.su

Vladimir A. Glushkov
Institute of Technical Chemistry, Perm, Russia
E-mail: ushka@newmail.ru

Anton V. Gulevich
Moscow State University, Department of Chemistry, Moscow, Russia
E-mail: antgul@mail.ru

Bodo Mario Gyuris
University of Szeged, Department of Medical Chemistry, Szeged, Hungary
E-mail: gyuris.mario@gmail.com

Azizollah Habibi
Tabiat Moallem University, Tehran, Iran
E-mail: azhabibi81@yahoo.com

Rahim Hekmatshoar
Department of Chemistry, School of Science, Alzahra University, Vanak, Tehran, Iran
E-mail: rhekmatu@yahoo.com

Prof. Majid M. Heravi
Department of Chemistry, School of Sciences, Azzahra University, Vanak, Tehran, Iran

Prof. Christopher Hulme
University of Arizona, Tucson, AZ, USA
E-mail: hulme@pharmacy.arizona.edu

Dr. Alexey Petrovich Ilyin
Chemical Diversity Research Institute, Khimki, Russia
E-mail: api@chemdiv.com

Yulia Sergeevna Isakova
Urals State Technical University-UPI, Ekaterinburg, Russia
E-mail: mcr2009@mail.ustu.ru

Kristina Igorevna Ivanova
Urals State Technical University-UPI, Ekaterinburg, Russia
E-mail: mcr2009@mail.ustu.ru

Dr. Maria Nikolaevna Ivantsova
Urals State Technical University-UPI, Ekaterinburg, Russia
E-mail: ivantsova@mail.ustu.ru

Prof. Laurent El Kaïm
Unité Chimie et Procédés, Ecole Nationale Supérieure de Techniques Avancées, Paris, France
E-mail: laurent.elkaim@ensta.fr

Dr. Cedric Kalinski
Priaxon AG, München, Germany
E-mail: kalinski@priaxon.com

Dr. Ivan Kanizsai
Avidin Biotechnology Ltd., Szeged, Hungary
E-mail: kanizsai.ivan@gmail.com

Dr. Alireza Kazemizadeh
Department of Chemistry, Islamic Azad University – Zanjan Branch, Zanjan, Iran
E-mail: alirezakazemizadeh@yahoo.com

Mohammad Khorassani
Department of Chemistry, School of Science, Alzahra University, Vanak, Tehran, Iran
E-mail: majid1313@yahoo.com

Nicola Kielland
Institute for Research in Biomedicine (IRB Barcelona), Barcelona, Spain
E-mail: nicola.kielland@irbbarcelona.org

Prof. Kyo Seon Kim
Kangwon National University, Department of Chemical Engineering, Chuncheon, Korea
E-mail: kkyoseon@kangwon.ac.kr

Prof. Sung Joon Kim
Kangwon National University, Department of Chemical Engineering, Chuncheon, Korea
E-mail: kkyoseon@kangwon.ac.kr

Anna Gennadievna Kombarova
Millab, Moscow, Russia
E-mail: ak@millab.ru

Andrei Nikolaevich Komogortsev
N.D. Zelinsky Institute of Organic Chemistry, Russian Academy of Sciences, Moscow, Russia
E-mail: dna5@mail.ru

Dr. Mikhail Krasavin
Chemical Diversity Research Institute, Khimki, Moscow Reg., Russia
E-mail: myk@chemdiv.com

Dr. Dmitry Vladimirivich Kravchenko
Chemical Diversity Research Institute, Khimki, Moscow Reg., Russia
E-mail: dk@chemrar.com

Alens Kriagh
IBC Corporate Travel, Ekaterinburg, Russia
E-mail: conf@ibc.ru

Yulia Sergeevna Kudyakova
I.Ya. Postovsky Institute of Organic Synthesis, Urals Branch of Russian Academy of Sciences, Ekaterinburg, Russia
E-mail: kud@ios.uran.ru

Dr. Christian Josef Kühnlein
Priaxon AG, München, Germany
E-mail: contact@priaxon.com

Dr. Volodymyr Kysil
ChemDiv. Inc., San Diego, CA, USA
E-mail: vkysil@chemdiv.com

Dr. Benedicte Lallemand
UCB Pharma S.A., Bruxelles, Belgium
E-mail: irene.ilkow@ucb.com

Prof. Grifols Rodolfo Lavilla
Institute for Research in Biomedicine, Barcelona Science Park, Barcelona, Spain
E-mail: rlavilla@pcb.ub.es

Tilman Lechel
Institut für Chemie und Biochemie, Freie Universität Berlin, Berlin, Germany
E-mail: tile@chemie.fu-berlin.de

Dr. Xiaoguang Lei
National Institute of Biological Science, Beijing, China
E-mail: leixiaoguang@nibs.ac.cn

Dr. Niels Lindner
Bayer Schering Pharma AG, Wuppertal, Germany
E-mail: niels.lindner@bayerhealthcare.com

Anna Vladimirovna Listratova
Russian Peoples' Friendship University, Moscow, Russia
E-mail: winterstory@yandex.ru

Ren Liuyi
Central South University, Changsha, China
E-mail: rly1015@yahoo.com.cn

Prof. Ping Lu
Chemistry Department, Zhejiang University, Hangzhou, China
E-mail: pinglu@zju.edu.cn

Prof. Andrey Nikolaevich Maslivets
Perm State University, Perm, Russia
E-mail: koh2@psu.ru

Prof. Anatoly Ivanovich Matern
Urals State Technical University-UPI, Ekaterinburg, Russia
E-mail: rector@mail.ustu.ru

Prof. Alevtina Sergeevna Medvedeva
A.E. Favorsky Irkutsk Institute of Chemistry, Siberian Branch of Russian Academy of Sciences, Irkusk, Russia
E-mail: amedved@irioch.irk.ru

Eugen Merkul
Institut für Organische Chemie und Makromolekulare Chemie, Heinrich-Heine-Universität Düsseldorf, Düsseldorf, Germany
E-mail: E.Merkul@uni-duesseldorf.de

Dr. Maxim Anatolievich Mironov
TOSLab, Ekaterinburg, Russia
E-mail: mironov@mail.ustu.ru

Riccardo Mossetti, M.S.
DiSCAFF, Università del Piemonte Orientale, Novara, Italy
E-mail: rickymoss@live.it

Prof. Thomas Jorma Johannes Müller
Heinrich-Heine-Universitaet Duesseldorf, Duesseldorf, Germany
E-mail: ThomasJJ.Mueller@uni-duesseldorf.de

Dr. Jamal Mohammad Reza Naimi
Organic Chemistry Research Laboratory, Department of Chemistry, Iran University of Science and Technology, Tehran, Iran
E-mail: naimi@iust.ac.ir

Dr. Farough Nasiri
University of Kurdistan, Sanandaj, Iran
E-mail: fnasiri@uok.ac.ir

Prof. Valentine Georgievich Nenajdenko
Moscow State University, Department of Chemistry, Moscow, Russia
E-mail: nen@acylium.chem.msu.ru

Oluwaseun Adedayo Olaitan
Cairo Wireless Institute, Down Town, Egypt
E-mail: cheunolaitan@yahoo.com

Prof. Michael George Organ
Department of Chemistry, York University, Toronto, ON, M3J 1P3, Canada
E-mail: organ@yorku.ca

Prof. Romano Vincenzo Antonio Orru
Department of Chemistry & Pharmaceutical Sciences, Vrije University, Amsterdam, The Netherlands
E-mail: orru@few.vu.nl

Prof. Stanislaw Ostrowski
Institute of Chemistry, University of Podlasie, Siedlce, Poland
E-mail: stan@ap.siedlce.pl

Dr. Shivaji Sandu Pandit
Department of Chemistry, Arts, Science and Commerce College, Kolhar, India
E-mail: akankshapandit2002@yahoo.com

Morejon Orlando Pando, M.S.
Leibniz Institut fur Planzenbiochemie, Halle, Germany
E-mail: omorejon@ipb-halle.de

Olga Panova
Department of Physical and Colloidal Chemistry, Perm State Pharmaceutical Academy, Perm, Russia
E-mail: panova.olga@gmail.com

Igor V. Paramonov
Urals State Technical University-UPI, Ekaterinburg, Russia
E-mail: beliaev@mail.ustu.ru

Olga Perevalova
Department of Physical and Colloidal Chemistry, Perm State Pharmaceutical Academy, Perm, Russia
E-mail: zakaz@ibc.ru

Dr. Tracey Pirali
DiSCAFF, Università degli Studi del Piemonte Orientale, Novara, Italy
E-mail: tracey.pirali@pharm.unipmn.it

Victor V. Potapov
Department of Organic Chemistry, Chemical Diversity Research Institute, Khimki, Moscow Reg., Russia,
E-mail: yai@chemdiv.com

Nikolai M. Przhevalski
Russian State Agrarian University – Moscow Timiryazev Agricultural Academy, Moscow, Russia
E-mail: prjevalski@mail.ru

Prof. Chuanguang Qin
Faculty of Life Science, Northwestern Polytechnical University, Xi'an, China
E-mail: qinchg@nwpu.edu.cn

Natalia Rassudihina
Perm State Pharmaceutical Academy, Perm, Russia
E-mail: limans@list.ru

Prof. Renata Riva
Università di Genova, Dipartimento di Chimica e Chimica Industriale, Genova, Italy
E-mail: riva@chimica.unige.it

Dr. Svetlana Alexandrovna Romanova
Sovay Pharma, Ekaterinburg, Russia
E-mail: rsa11@yandex.ru

Dr. Igor Borisovich Rozentsveig
A.E. Favorsky Irkutsk Institute of Chemistry, Siberian Branch of Russian Academy of Sciences, Irkusk, Russia
E-mail: i_roz@irioch.irk.ru

Yulia Sergeevna Rozhkova
Institute of Technical Chemistry of Urals Branch of Russian Academy of Sciences, Perm, Russia
E-mail: rjs@mail.ru

Prof. Vladimir Leonidovich Rusinov
Enamine Ltd., Kiev, Ukraine
E-mail: rusinov@mail.ustu.ru

Dr. Sergey V. Ryabukhin
Enamine Ltd., Kiev, Ukraine
E-mail: ryabukhin@mail.enamine.net

Dr. Elena Vladimirovna Sadchikova
Urals State Technical University-UPI, Ekaterinburg, Russia
E-mail: selena-ekb@rambler.ru

Dr. Eva Seip
Elsevier MDL GmbH, Germany
E-mail: E.Seip@elsevier.com

Prof. Ahmad Shaabani
Department of Chemistry, Shahid Beheshti University, Tehran, Iran
E-mail: a-shaabani@cc.sbu.ac.ir

Marina Shatalina
IBC Corporate Travel, Ekaterinburg, Russia
E-mail: events@ibc.ru

Prof. Hassan Sheibani
Department of Chemistry, Shahid Bahonar University of Kerman, Kerman, Iran
E-mail: hsheibani@mail.uk.ac.ir

Prof. Masakatsu Shibasaki
Graduate School of Pharmaceutical Sciences, University of Tokyo,
Hongo 7-3-1, Bunkyoku, Tokyo 113-0033, Japan
E-mail: ysuya@mol.f.u-tokyo.ac.jp

Olga Vladimirovna Shilova
Chemical Diversity Research Institute, Khimki, Moscow Reg., Russia
E-mail: sov@iihr.ru

Alexandre M. Shkirando
Department of Organic Chemistry, Chemical Diversity Research Institute,
Khimki, Moscow Reg., Russia,
E-mail: yai@chemdiv.com

Prof. Yurii Vladimirovoch Shkylaev
Institute of Technical Chemistry of Urals Branch of Russian Academy
of Sciences, Perm, Russia
E-mail: yushka@newmail.ru

Dr. Ivan Vladimirovich Smolyaninov
Southern Scientific Centre of Russian Academy of Sciences, Russia
E-mail: thiophen@mail.ru

Prof. Cao Song
Shanghai Key Laboratory of Chemical Biology, Center of Fluorine Chemical
Technology, School of Pharmacy, East China University of Science and
Technology, Shanghai, China
E-mail: scao@ecust.edu.cn

Dr. Eddy Sotelo
University of Santiago de Compostela, Santiago de Compostela, Spain
E-mail: eddy.sotelo@usc.es

Julia Alekseevna Titova
I.Ya. Postovsky Institute of Organic Synthesis, Urals Branch of Russian Academy
of Sciences, Ekaterinburg, Russia
E-mail: titova@ios.uran.ru

Dr. Maria Igorevna Tokareva
Urals State Technical University-UPI, Ekaterinburg, Russia
E-mail: tokarevami@mail.ru

Prof. Yury Vasil'evich Tomilov
N.D. Zelinsky Institute of Organic Chemistry, Russian Academy of Sciences, Moscow, Russia
E-mail: tom@ioc.ac.ru

Dr. Gian Cesare Tron
Università del Piemonte Orientale, DISCAFF, Novara, Italy
E-mail: tron@pharm.unipmn.it

Dr. Igor Alekseevich Ushakov
A.E. Favorsky Institute of Chemistry, Siberian Branch of the Russian Academy of Sciences, Irkutsk, Russia
E-mail: nmr@istu.edu

Dr. Oleg Gennadievich Utkin
Elsevier, Russia
E-mail: O.Utkin@elsevier.com

Buchi Reddy Vaddula
Birla Institute of Technology and Science, Pilani, India
E-mail: buchi639@yahoo.co.in

Prof. Erik Van der Eycken
Laboratory for Organic & Microwave-Assisted Chemistry (LOMAC), Katholieke Universiteit Leuven (K.U. Leuven), Leuven, Belgium
E-mail: erik.vandereycken@chem.kuleuven.be

Viktoria Nikolaevna Vichegjanina
Perm State Pharmaceutical Academy, Perm, Russia
E-mail: aleksejvagapov@yandex.ru; perm@pfa.ru

Ivan Nikolaevich Vladimirov
Perm Academy of Pharmacy, Perm, Russia
E-mail: ivavladimirov@ya.ru

Alexandra Sergeevna Vorobyeva
Chemical Diversity Research Institute, Khimki, Moscow Reg., Russia
E-mail: vas@iihr.ru

Dr. Leonid Gennadievich Voskressensky
Russian Peoples' Friendship University, Moscow, Russia
E-mail: lvoskressensky@sci.pfu.edu.ru

Prof. Yanguang Wang
Department of Chemistry, Zhejiang University, Hangzhou, China
E-mail: orgwyg@zju.edu.cn

Prof. Ludger A. Wessjohann
Leibniz Institute of Plant Biochemistry, Halle (Saale), Germany
E-mail: wessjohann@ipb-halle.de

Prof. Grigory Yakovlev
Izhevsk State Technical University, Izhevsk, Russia
E-mail: jakowlew@udm.net

Prof. Bing Yan
Shandong University, Shandong, P.R. China
E-mail: dr.bingyan@gmail.com

Prof. Miguel Yus
Institute for Organic Synthesis and Department of Organic Chemistry, Faculty of Science, University of Alicante, Apdo.99, 03080 Alicante, Spain
E-mail: yus@ua.es

Aliaksandr Zhdanko
Moscow State University, Moscow, Russia
E-mail: vinceeero@gmail.com

Dr. Maria Zhidovinova
Ghent University, Belgium
E-mail: zhidovinova@yandex.ru

Prof. Jieping Zhu
Institut de Chimie des Substances Naturelles, CNRS, Gif-sur-Yvette Cedex, France
E-mail: zhu@icsn.cnrs-gif.fr

Index